Language as a Scientific Tool

Language is the most essential medium of scientific activity. Many historians, sociologists, and science studies scholars have investigated scientific language for this reason, but few have examined those cases where language itself has become an object of scientific discussion. Over the centuries scientists have sought to control, refine, and engineer language for various epistemological, communicative, and nationalistic purposes. This book seeks to explore cases in the history of science in which questions or concerns with language have bubbled to the surface in scientific discourse. This opens a window into the particular ways in which scientists have conceived of and construed language as the central medium of their activity across different cultural contexts and places and the clashes and tensions that have manifested their many attempts to engineer it to both preserve and enrich its function. The subject of language draws out many topics that have mostly been neglected in the history of science, such as the connection between the emergence of national languages and the development of science within national settings, and allows us to connect together historical episodes from many understudied cultural and linguistic venues such as Eastern European and medieval Hebrew science.

Miles MacLeod is Professor for Philosophy of Science at the University of Twente, The Netherlands.

Rocío G. Sumillera is Assistant Professor in the Department of English and German, University of Granada, Spain.

Jan Surman is Postdoctoral Research Associate at the Leibniz Graduate School at the Herder Institute for Historical Research on East Central Europe in Marburg.

Ekaterina Smirnova is currently affiliated with Sciences Po (Paris) and the STS Center in EUSP.

Routledge Studies in Cultural History

For a full list of titles in this series, please visit www.routledge.com

17 **Politics of Memory**
Making Slavery Visible in the Public Space
Edited by Ana Lucia Araujo

18 **Neutrality in Twentieth-Century Europe**
Intersections of Science, Culture, and Politics after the First World War
Edited by Rebecka Lettevall, Geert Somsen, and Sven Widmalm

19 **Americans Experience Russia**
Encountering the Enigma, 1917 to the Present
Edited by Choi Chatterjee and Beth Holmgren

20 **A Social History of Disability in the Middle Ages**
Cultural Considerations of Physical Impairment
Irina Metzler

21 **Race, Science, and the Nation**
Reconstructing the Ancient Past in Britain, France, and Germany
Chris Manias

22 **Identity, Aesthetics, and Sound in the Fin de Siècle**
Redesigning Perception
Dariusz Gafijczuk

23 **Disease and Crime**
A History of Social Pathologies and the New Politics of Health
Edited by Robert Peckham

24 **Critical Perspectives on Colonialism**
Writing the Empire from Below
Edited by Fiona Paisley and Kirsty Reid

25 **Old World Empires**
Cultures of Power and Governance in Eurasia
Ilhan Niaz

26 **The Afterlife of Used Things**
Recycling in the Long Eighteenth Century
Edited by Ariane Fennetaux, Amélie Junqua, and Sophie Vasset

27 **Holocaust Consciousness in Contemporary Britain**
Andy Pearce

28 **The Invention of Race**
Scientific and Popular Representations
Edited by Nicolas Bancel, Thomas David, and Dominic Thomas

29 **Indigenous Networks**
Mobility, Connections, and Exchange
Edited by Jane Carey and Jane Lydon

30 **Shadows of the Slave Past**
Memory, Heritage, and Slavery
Ana Lucia Araujo

31 **Expedition into Empire**
Exploratory Journeys and the Making of the Modern World
Edited by Martin Thomas

32 **Luxury and Gender in European Towns, 1700–1914**
Edited by Deborah Simonton, Marjo Kaartinen, and Anne Montenach

33 **Reassessing the Transnational Turn**
Scales of Analysis in Anarchist and Syndicalist Studies
Edited by Constance Bantman and Bert Altena

34 **Conciliation on Colonial Frontiers**
Conflict, Performance, and Commemoration in Australia and the Pacific Rim
Edited by Kate Darian-Smith and Penelope Edmonds

35 **Scotland and the Caribbean, c.1740–1833**
Atlantic Archipelagos
Michael Morris

36 **Case Studies and the Dissemination of Knowledge**
Edited by Joy Damousi, Birgit Lang, and Katie Sutton

37 **Visualizing Jews Through the Ages**
Literary and Material Representations of Jewishness and Judaism
Edited by Hannah Ewence and Helen Spurling

38 **Higher Education and the Growth of Knowledge**
A Historical Outline of Aims and Tensions
Michael Segre

39 **New Perspectives on Russian-American Relations**
Edited by William Benton Whisenhunt and Norman E. Saul

40 **The Holocaust in the Twenty-First Century**
Contesting/Contested Memories
Edited by David M. Seymour and Mercedes Camino

41 **Jesuits at the Margins**
Missions and Missionaries in the Marianas (1668–1769)
Alexandre Coello de la Rosa

42 **Travelling Notions of Culture in Early Nineteenth-Century Europe**
Edited by Hannu Salmi, Asko Nivala, and Jukka Sarjala

43 **Language as a Scientific Tool**
Shaping Scientific Language Across Time and National Tradition
Edited by Miles MacLeod, Rocío G. Sumillera, Jan Surman, and Ekaterina Smirnova

Language as a Scientific Tool

Shaping Scientific Language Across Time and National Tradition

Edited by Miles MacLeod, Rocío G. Sumillera, Jan Surman, and Ekaterina Smirnova

Routledge
Taylor & Francis Group
LONDON AND NEW YORK

First published 2016 by Routledge

2 Park Square, Milton Park, Abingdon, Oxfordshire OX14 4RN
52 Vanderbilt Avenue, New York, NY 10017

Routledge is an imprint of the Taylor & Francis Group, an informa business

First issued in paperback 2019

Library of Congress Cataloging-in-Publication Data
Names: MacLeod, Miles, editor.
Title: Language as a scientific tool : shaping scientific language across time and national tradition / edited by Miles MacLeod [and three others].
Description: New York : Routledge, 2016. | Series: Routledge studies in cultural history ; 43 | Includes bibliographical references and index.
Identifiers: LCCN 2015041763 | ISBN 9781138101050 (alk. paper)
Subjects: LCSH: Communication in science. | Science—Language. | Language, Universal.
Classification: LCC Q223 .L264 2016 | DDC 501/.4—dc23
LC record available at http://lccn.loc.gov/2015041763

ISBN: 978-1-138-10105-0 (hbk)
ISBN: 978-0-367-26392-8 (pbk)

Typeset in Sabon
by Apex CoVantage, LLC

Contents

PART 2
The Creation of Scientific Terminology

PART 3
Imagining Universal Languages

Figures

Tables

Acknowledgments

We would like to thank the Austrian Academy of Sciences and especially Dr. Johannes Feichtinger for funding and helping organize the initial workshop that led to the development of this book, as well as John Kriege for his advice on initial versions of this volume. We would also like thank the anonymous reviewers for their comments.

Miles MacLeod's participation in this project was supported by postdoctoral fellowships at the Academy of Finland Centre of Excellence for the Philosophy of the Social Sciences and at the Konrad Lorenz Institute for Evolution and Cognition Research (Altenberg, Austria) and by a US National Science Foundation grant: number DRL097394084. Rocío G. Sumillera would like to acknowledge the funding from the research project of the Universitat de València 'Construyendo Europa: Literaturas en Contacto y Arquetipos Literarios' (UV-UNV-PRECOMP14-206579).

1 Introduction

Miles MacLeod, Rocío G. Sumillera, Jan Surman, and Ekaterina Smirnova

It goes without saying that language is the central medium in which science operates. However, it is not the only medium of science. Science takes place in material practices and processes; it takes place in spaces like laboratories and computer labs, it communicates and constructs through visual images and graphical representations. But the fundamental part of language in communicating and recording information, and in theory-building itself, has rightly attracted a spectrum of philosophers, historians, sociologists, and anthropologists of science interested in language. These scholars have been primarily interested in how the cultural constraints of language have informed and determined the practice of science, fueled scientific controversies, and governed the shape and content of scientific theories.[1] One might see little room for yet another volume that explores how the language in which scientists have operated and communicated reflects or embodies a scientific culture, a culture which in turn reflects or embodies wider political and socio-cultural practices, relations, and attitudes. Nor would there seem room for yet another volume on how scientists employ rhetoric and linguistic devices to convince or cajole others to their point of view. In fact we do not provide a discussion about scientific rhetoric, although we will not deny that much of what we have documented has rhetorical function and could be studied as such.

What our collection of texts is interested in is somewhat different. As much as language is a medium of science, it is also at times an object of active scientific reflection and manipulation. Scientists and natural philosophers have a history of discussing, theorizing about, and restructuring language in science.[2] This history traces back to ancient discussions about nature and culture, through medieval discussions on instrumentation, and into the earliest sixteenth and seventeenth centuries concerns with specialized natural historical and experimental language, universal languages, and mathematics as the language of the heavens.[3] It might even be said that as soon as anything that can be roughly considered "science" existed there have also been discussions and debates on language, if only on the matter of scientific terminology—one only has to think of the frequent discussions amongst scientists of various nations or ethnic groups regarding how to vernacularize science into a particular national or ethnic language. And certainly, terminology planning commissions—both national and international—have played an integral part in driving scientific thought about language at least since the late nineteenth century. Finally, the recent advent of computation has turned scientific attention to questions regarding the relationships between natural languages and computational or mathematical languages.[4] At the same time, the issue of how to translate scientific writings from a natural language to another, and thus facilitate the transfer of knowledge, constituted a parallel, although not equal, source of questions on how to model language to enable flexibility and creativity for generating new terminology and enable intercultural communication at the same time. Here we are in company with feminist and postcolonial studies, which have recently begun to explore the broader societal role of scientific language as a means of controlling cultural hegemony.[5]

We are interested in the historical and contemporary episodes in which scientific groups have taken up the subject of language themselves and attempted to reform or restructure it to achieve what has turned out to be a wide range of instrumental goals. For these episodes we find the tool metaphor most apt, since in the cases we discuss, natural philosophers and scientists are concerned with how to shape, manipulate, and define either language as a whole or particular languages in order to achieve a certain end, and in many instances this has afforded them philosophical claims about language and its role in science and society more broadly.

In more concrete terms *Language as a Scientific Tool* considers cases in which researchers have objectified the linguistic medium in which they ordinarily operate, attempting to analyze it and ultimately control or reform it in some way. We study the means by which these processes of objectification and manipulation have taken place, but also the variety of ends that these linguistic restructuring projects have been put to and the constraints governing them. These ends are diverse and reflect the cultural and social contexts in which they are historically situated. They have included epistemic ends that aim to define and construct scientific language in a way that best reflects a set of historically contingent epistemic values. Such values express how language needs to be understood and applied to best obtain and report information about the world. They have historically included preferences and claims for clarity, tractability, and communicability but also translatability and more recently computability.[6] These ends have also included desires to use ideas and concepts of language to impose or promote an ontological perspective that privileges certain languages like the languages of mathematics and physics. In this respect, attempts have been made historically to portray language as out there in the universe, or at least to centralize the role of the metaphor of language in representing the basic organization of nature. Finally, much of the scientific concern with language has been driven by nationalistic ends as scientists have sought to give their own linguistic groups access to knowledge, but also to gain nationalistic control over its production through processes of vernacularization.[7]

As such we are concerned with a plethora of issues related by the common disposition of scientists and natural philosophers to treat language as a tool. We map some of the different historical and cultural contexts in which this disposition has taken shape, and discuss how these contexts have driven particular interactions with the subject of language or individual languages, and the particular structures and properties of languages and their evolution in these contexts. Themes which bubble to the surface as a result of these case studies include the relationship between general or natural languages and the subcategories of scientific languages; the connection between the emergence of national languages and the development of science within national settings; and the relations between the creation of scientific languages and the appearance of new scientific disciplines. Of particular interest is the repeated tension between the constraints of everyday languages and entrenched scientific languages, with the social and cultural contexts they embody, and the desire of scientists to manipulate and deploy language to particular ends. These have frequently collided in scientific processes of reforming language, requiring compromise and the negotiation by scientists of thickets of meaning and the resistance of language to control. *Language as a Scientific Tool* is the first volume to explicitly tackle these questions across a range of historical and cultural contexts.

In particular we pay attention to science in languages other than English, bringing together scholars working on the historical reception of science and the modern production of science in European and Eastern European languages. We have obtained perspectives on these questions of language from places and historical contexts that are under-represented in contemporary history, sociology, and philosophy of science but that are necessary to having any broad scope of understanding of how language has been manipulated and reformed in different cultural environments and what different ends have informed this process.

The volume is arranged in an opening essay and three parts, each unfolding a different approach to the complex relationship between language and science. In his opening essay, Matthias Dörries understands the emergence of modern physics as inseparable from the practices and knowledge of languages. While the statement that modern science in Europe evolved from Renaissance culture and its study of ancient texts and languages is a commonplace in standard histories of science, Dörries argues that historians have only recently started to study systematically the linguistic and literary environments that nurtured new ideas in early modern natural philosophy. His essay compares ways in which the research of three physicists (Galileo, Heinrich Kayser, and Werner Heisenberg) referred to language in their scientific research, proving that their reflection on language remarkably intertwined with their scientific work, rather than merely constituting marginal and anecdotal concerns.

Part 1 of the volume addresses some of the discourses and contexts in which language itself has become a participant and taken on functional roles and how specific problems came to be viewed as problems of language or problems language should resolve. In the first place we are interested in the historical links that track the emergence of one of the principal ways in which language is considered a tool for much of modern science, namely, as an arbitrary symbolic system of representation. This occurs during the period usually labeled the Enlightenment and is tied up in the Western context with anti-Aristotelian philosophical movements. Today, the idea that language is an arbitrary tool rather than a resource of intrinsic facts about the world is a universal attitude. It is a given for most scientific enterprises except perhaps physics and mathematics, as mathematics has a long history of being treated as a language with ontological significance. As Miles MacLeod explains, this attitude was not one that came down from posterity to natural philosophers of the seventeenth century, but had to be constructed. Those most active in this construction were the British Empiricists, who traced their ideas to Francis Bacon's early polemics against the "idols of the marketplace". Indeed this reconstruction of language as arbitrary and free of any inherent meaning about the world or the universe served many contextual purposes such as rhetorically undermining competitors like scholasticism, undergirding specific epistemic and ontological values of experimental philosophy and natural history. The reconstruction of language as a tool at the hands of people like Locke was a deliberated construction. Whatever scientists think obvious today was not necessarily obvious at the time and had to be justified in complex ways and articulated to serve various argumentative ends.

The seventeenth century also constitutes the starting point of the essay by Jerzy Biniewicz, who changes the geographical coordinates of his study from Britain to Poland. His essay analyzes the first scientific and scientific-didactic texts that appeared in the sixteenth and seventeenth centuries with a view to understanding the mechanism of forming a scientific picture of the world that differed from popular knowledge. Language hence appears as an instrument of thinking, description, and communication that enables encoding, describing, and communicating the world. Martin Herrnstadt and Laurens Schlicht's essay connects discussions on language with those on history in late eighteenth- and early nineteenth-century France, a time when language was situated in the complex intersections between nature, the savage, and civilization and interest in the "real" savage or the "real" native grew, as the writings of Roch Ambroise Sicard, Jean Itard, or Joseph-Marie Degérando illustrate. This interest is inseparable from a specific view on what history is and from a larger willingness to explore the "real" nature of language, which was thought should conform to the universal laws behind reason. By focusing on the figures of the savage and the deaf-mute, Schlicht demonstrates how language became simultaneously a tool and an object of research and how research on the roots of language was paradoxically based on empirically studying individuals without any language at all. Finally, Priya Venkatesan Hays shows that when assessing question of uncertainty in scientific language, politicians and scientists seem to be on opposite ends of the epistemological spectrum. Politicians would like for scientists to caveat their claims, while amongst scientists the opposite occurs. Venkatesan Hays's essay addresses how scientists mediate the uncertainty and ambiguity of language to manage the rhetorical effects of language in their work.

Part 2 deals with the creation of scientific terminology through four case studies with different national backgrounds and linguistic combinations. The first essay is by Josefina Rodríguez Arribas, who focuses on three twelfth-century treatises in Hebrew by Abraham ibn Ezra devoted to the description and explanation of astrolabes. These treatises, which are amongst the earliest manifestations of scientific writing in Hebrew, are a means to research on the process of specialization and creation of technical terms in Hebrew in the Iberian Peninsula. Rodríguez Arribas argues that the scientific writings of authors such as Abraham ibn Ezra made medieval Hebrew a language capable of communicating science and of scientific research. The second case study investigates a particular transformation of the concept of *opyt* in Russian language at the verge of modernity. Ekaterina Smirnova explains that in Russian the term *opyt* designates two different concepts which in most European languages are expressed by different words: the concept of knowledge and skills and the concept of specially conducted and controlled procedures of testing and examination (in English, *experience* and *experiment*, respectively). The essay concentrates on this particular example to show how the appropriation and further usage of scientific concepts from other languages is influenced and constrained by the linguistic and cultural contexts of the translating language.

Third, Jan Surman traces the interrelations between science, language, and nationalism in nineteenth-century chemistry by focusing on the history of the term *oxygène*. Surman claims that the puristic renaming of the term in Danish and Polish can be understood as an intersection between scientific theory, an applied concept of scientific language, and the proposed ideal of national language. When the French Morveau-Lavoisier nomenclature system was established, it treated chemistry as a linguistic system, which opened a way for different appropriations of the terminology. As the terms were to denote "unique" features of their chemical objects, in due course the development of chemistry challenged traditional nomenclature, creating an opportunity to re-name compounds whose names were "false". The phenomenon of name replacing occurred in Danish and Polish during the discussion about vernacularization of the language of science. While Lavoisier intended to make his terms "abstract" by choosing Greek forms, scholars opting for the vernacular were resolved to make science more understandable (as they put it), which led to different outcomes and generated new issues. In the fourth case study, Helena Durnová discusses the role of language in the history of computing technology: in the 1950s, the terminology in computing technology was still unstable, and different terms—loaded with ideological implications—were chosen by different groups. Concentrating on the use of language in the computing technology communities in the Soviet bloc, in particular in Czechoslovakia, this essay maps the use of various terms and discusses the possible strategic choices scientists faced in their construction of computational languages. Special emphasis is placed on those scientists concerned with translation into their own mother tongue and on how their terminology mirrored their belief in the power of the new technology.

Part 3 approaches from different perspectives the historically recurring issue of universal languages and their implication for scientific development. Rocío G. Sumillera's essay focuses on the interest in the seventeenth century in the creation of a universal language that would overcome the obstacles of the multiplicity of tongues, as well as solve the arbitrary relationship between things and words. Attention is particularly drawn to John Wilkins's *An Essay towards a Real Character, and a Philosophical Language* (1668) and to its impact on the literary imagination of Jonathan Swift: in *Gulliver's Travels* (1726), Swift generally mocks the linguistic projects of the Royal Society, including the invention of a universal and philosophical language. Notwithstanding the apparent opposition between Wilkins's and Swift's linguistic ideas, both authors ultimately share the conviction that linguistic mutability is undesirable whereas a fixed language is the ideal. A different universal language project is analyzed by Markus Krajewski: the standardized auxiliary language designed by laymen such as Ludwik Lejzer Zamenhof or Johann Martin Schleyer and advocated by renowned European scholars such as Louis Couturat and Wilhelm Ostwald around 1900. Because the project of the World Auxiliary Language Movement—which followed the example of Ido, a derivative of Esperanto—coincided with the beginning of World War I, the simplified globality the linguistic scheme aimed to establish was drastically interrupted.

To close the volume, Scott L. Montgomery reflects on the fact that English is currently the first world-scale language in the natural sciences, on the impact of a global language on science in the present, and on the limits and drawbacks of a powerful lingua franca gaining authority. Montgomery sets out to answer whether there are important disadvantages that stand out, and, if so, how serious they are both for the present and the future. Given the many years of training and the intense competition for resources and rewards in contemporary science, Montgomery postulates different ways to address such problems. Questions about fairness, access to knowledge, and linguistic diversity, for instance, are nowadays regularly posed, and this chapter precisely tackles central issues for the future of scientific endeavor, in terms of researchers themselves and their social practices.

One of the conclusions emerging from these essays is that language, the indispensable tool for science, is able to reflect, represent, and communicate the complexities and changing thoughts of scientific imagination and inventiveness thanks to its ever-changing nature and malleability. Not only does language allow scientists to get their messages across, communicate their experiments, and share their conclusions, but it plays a key role in the process of shaping scientific ideas. Language is thus not only an essential vehicle for science, but a force that shapes the production of science itself by molding or structuring the way our minds operate. Far from being a passive conductor of science, language is both an active tool for science and the context in which science itself develops. As the following essays illustrate, language is a restless agent in scientific production which inevitably intertwines with science in complex webs of political, economical, and socio-cultural nature that have determined their common history.

NOTES

1. For some of the more recent overviews, see Leah Ceccarelli, *Shaping Science with Rhetorics. The Case of Dobzhansky, Schrodinger and Wilson* (Chicago and London: University of Chicago Press, 2001); Michael A.K. Halliday, *The Language of Science* (London: Continuum, 2004); Roy Harris, *The Semantics of Science* (London, New York: Continuum, 2005); Alan G. Gross, *Starring the Text: The Place of Rhetoric in Science Studies* (Carbondale: Southern Illinois University Press, 2006); Jeanne Fahnestock, "The Rhetoric of Natural Sciences," in *The SAGE Handbook of Rhetorical Studies*, eds. Andrea A. Lunsford, Kirt H. Wilson and Rosa A. Eberly (Thousand Oaks: Sage 2009, 175–191); Randy Harris, "Alan Gross and the Rhetoric of Science," *Perspectives on Science* 17.3 (2009): 346–380.
2. See Bruno Snell, "The Forging of a Language for Science in Ancient Greece," *The Classical Journal* 56.2 (1960): 50–60; Matthias Dörries, "Language as a Tool in Science," in *Experimenting in Tongues. Studies in Science and Language* (Stanford: Stanford University Press, 2002, 1–20); Ann Moss, *Renaissance Truth and the Latin Language Turn* (Oxford and New York: Oxford University Press, 2003); Michle Goyens, Pieter De Leemans, and An Smets, eds. *Science Translated: Latin and Vernacular Translations of Scientific Treatises in Medieval Europe* (Leuven: Leuven University Press, 2008).
3. On the changes in language ideology, see James Joseph Bono, *The Word of God and the Languages of Man: Interpreting Nature in Early Modern Science and Medicine. Vol. 1. Ficino to Descartes* (Madison: University of Wisconsin Press, 1995); Jürgen Schiewe, *Die Macht der Sprache. Eine Geschichte der Sprachkritik von der Antike bis zur Gegenwart* (München: C. H. Beck, 1998). On the changing role of language in science, see Christian Licoppe, *La formation de la pratique scientifique* (Paris: La Decouverte, 1998), and on interdisciplinary appropriation of language, Elisabeth Garber, *The Language of Physics. The Calculus and the Development of Theoretical Physics in Europe, 1750–1914* (New York: Birkhäuser, 1999).
4. Jörg Pflüger, "Language in Computing," in *Experimenting in Tongues*, ed. Matthias Dörries (Stanford: Stanford University Press, 2002, 125–162); see also chapter by Helena Durnová in this volume.
5. These recent discussions are far from over. Michael Lackner, Iwo Amelung, and Joachim Kurtz, eds. *New Terms for New Ideas. Western Knowledge and Lexical Change in Late Imperial China* (Leiden: Brill, 2001); Scott L. Montgomery, *Science in Translation: Movements of Knowledge through Cultures and Time* (Chicago: University of Chicago Press, 2002). See also Ulrich Ammon, *Ist Deutsch noch internationale Wissenschaftssprache?* (Berlin and New York: de Gruyter, 1999); Charles Durand, *La mise en place des monopoles du savoir* (Paris: Harmattan, 2001); Franz Nies, *Europa denkt mehrsprachig / L'Europe*

pense en plusieurs langues (Tübingen: Narr, 2005). For postcolonial and feminist criticism, see Donna Haraway, *Primate Visions* (New York: Routledge, 1989); Evelyn Fox Keller, *Secrets of Life, Secrets of Death: Essays on Language, Gender, and Science* (New York: Routledge, 1992); Kwasi Wiredu, *Conceptual Decolonization in African Philosophy: Four Essays* (Ibadan: Hope Publications, 1995); Dipesh Chakrabarty, *Provincializing Europe: Postcolonial Thought and Historical Difference* (Princeton: Princeton University Press, 2000).

6. Steven Shapin, "Pump and Circumstance: Robert Boyle's Literary Technology," *Social Studies of Science* 14.4 (1984): 481–520; Stephen Shapin and Simon Schäffer, *Leviathan and the Air-Pump: Hobbes, Boyle, and the Experimental Life* (Princeton: Princeton University Press, 1985); Pietro Corsi, "After the Revolution: Scientific Language and French Politics, 1795–1802," in *The Practice of Reform in Health, Medicine, and Science, 1500–2000*, eds. Margaret Pelling and Scott Mandelbrote (Aldershot: Ashgate, 2005, 223–245); Maurice Crosland, *The Language of Science: From the Vernacular to the Technical* (Cambridge: Lutterworth Press, 2006); Richard Rorty, *Philosophy and the Mirror of Nature* (Princeton: Princeton University Press, 2009).
7. Wolfgang Walter Menzel, *Vernakuläre Wissenschaft. Christian Wolffs Bedeutung für die Herausbildung und Durchsetzung des Deutschen als Wissenschaftssprache* (Tübingen: Niemeyer, 1996); Olga A. Valkova, "Wissenschaftssprache und Nationalsprache. Konflikte unter russischen Naturwissenschaftlern in der Mitte des 19. Jahrhunderts", in *Wissenschaft und Nation in der europäischen Geschichte, Wissenschaft und Nation in der europäischen Geschichte*, eds. Ralph Jessen and Jakob Vogel (Frankfurt am Main: Campus Verlag, 2002, 59–79); Anne Helga Hannesdóttir, "From Vernacular to National Language: Language Planning and the Discourse of Science in Eighteen Century Sweden", in *Languages of Science in the Eighteenth Century*, ed. Britt-Louise Gunnarson (Berlin and Boston: De Gruyters, 2011, 107–122); Jan Surman, "Science and Its Publics. Internationality and National Languages in Central Europe", in *The Nationalization of Scientific Knowledge in the Habsburg Empire, 1848–1918*, eds. Mitchell G. Ash and Jan Surman (Basingstoke: Palgrave Macmillan, 2012, 30–56).

2 Modern Science and the Spirit of Language, Literature, and Philology

Matthias Dörries

The emergence of modern science is inseparable from the study of languages. While the statement that modern science in Europe evolved from Renaissance culture and its examination of ancient texts and languages is commonplace in standard histories of science, historians have only recently started to study systematically how linguistic and literary environments actually nurtured strikingly new scientific ideas in early modern European natural philosophy and beyond.[1] Two key figures in the history of science, Galileo Galilei (1564–1642) and Werner Heisenberg (1901–1976), sharpened their minds by reflecting upon language consistently throughout their lives. This study led them to identify new alleys of research, refine their analysis and interpretation of new or unfamiliar objects and topics in astronomy or physics, and attribute sense to results that lay far outside the usual tracks of inquiry. Communicating their disturbing scientific results both to a scientific community and to society at large was essential to each, and they both put great effort into the framing and presentation of their scientific work to larger audiences. The very act of communication, in conversation with colleagues, carefully elaborated texts, or public presentations, gave shape to yet unspoken ideas, molded them to better fit the culture and styles of their time, and furthered their dissemination and ultimately their acceptance and confirmation. Both Galileo and Heisenberg were well prepared for this delicate task: they had enjoyed a broad education, covering the whole spectrum from humanistic training on the one side to philosophy of nature (or natural sciences), astronomy, and mathematics on the other.

For Galileo and Heisenberg, language served as a vehicle that mediated between their individual life-world experience and the culture in which they moved and worked. Modern science evolved out of a humanistic European culture, but developed in a tension with it. Mathematical formalism collided with a common language that only insufficiently caught what was expressed in abstract quantitative terms. Given these difficulties of expression and communication, the German philosopher Edmund Husserl in the 1930s identified a crisis in European culture, and turned to the *history* of mathematics and geometry, in the hope of bringing modern science back into European culture under the stewardship of philosophy.[2] Husserl's ambitious effort to place philosophy again at the top of human knowledge failed to convince scientists and others. However, scientists like Heisenberg faced the same problem Husserl did, and they took on the task of making sense of their work within their own culture.

In this essay, I will examine how three scientists engaged and struggled with language and literature. I will restrict myself to three case studies and rely heavily on two recent in-depth studies on Galileo and Werner Heisenberg, written by John L. Heilbron and Cathryn Carson, respectively.[3] Galileo will serve as a synecdoche of the revolutionary physics and astronomy of seventeenth-century Italy, while Heisenberg represents the revolutionary theoretical physics of twentieth-century Germany. I will contrast and compare their work and analysis with Heinrich Kayser (1853–1940), subject of my own previous study, a prototype of a late nineteenth-century classical physicist during the German *Kaiserreich*.[4] These three case studies show how thinking about language can intervene at any stage of the process of scientific discovery and how these interventions do not follow any prescribed paths or roles. Language is too complex and malleable and easily escapes any narrow definition or strict use. Acknowledging this complexity does not mean the defeat of analysis, but rather hints at language's constructive power for reasoning and world-disclosure and its capacity for building bridges between the known and the unknown, the familiar and the unfamiliar. Following Wittgenstein, one may speak of an ongoing linguistic game for expressing unspoken ideas.[5] There is this subtle game of using language pragmatically as a tool, all the while being guided by language's unlimited potentials.

1. GALILEO AS A LITERARY AND LINGUISTIC CRITIC

There is one Galileo, the prototype of the modern physicist, who did experiments, developed instruments, and searched for quantitative relationships in nature, not causal explanations. Then there is another Galileo, who was fascinated with astrology, disregarded Kepler's elliptical astronomy, and stuck to circles as expressions of beauty and perfection. Then there is yet another Galileo, who studied Latin, learned drawing at the design school of Pisa, played and studied music with his father, wrote poems, sketched the outlines of a play, engaged in debates about who was the greatest poet (Homer or Virgil), and commented on the literature of his time (Ariosto and Tasso). In *Galileo*, the historian of science John L. Heilbron firmly situates Galileo in the literary and linguistic framework of his time. The book explores how Galileo's new and striking scientific discoveries were made by a man originally deeply immersed in Aristotelian thought and the culture of his time—"a humanist of the old school".[6] The Galileo we celebrate today, the one who pointed his telescope at the moon and the planets and published his results in *Sidereus Nuncius* in 1610, made these discoveries only late in his life, in his mid-forties. Galileo's earlier career had shown promise in mathematics, to which he turned after studying medicine, ultimately becoming a professor of mathematics at the University of Pisa in 1589. However, the younger Galileo preferred literary circles to mathematical ones. Galileo cared about language and literature, giving lectures on literary topics and engaging in debates about the literary qualities of such eminent Italian writers as Dante, Ariosto, and Tasso. Here too Galileo made a substantial contribution to European and Italian culture: his commentary on Tasso is still required reading in Italian schools. None of these activities would support Galileo financially, but they shaped his thinking and his horizon. Galileo sharpened his mind and language by putting literary works under intense scrutiny and examination. Galileo may be best described as a "critic" with strong and clear-cut opinions of a Manichean character.[7]

Galileo's critical reading of Ariosto and Tasso provides insights into how these two works left a mark and how they inspired his future revolutionary scientific work and his literary presentation of it. For Heilbron, Ariosto's *Orlando furioso*, of the first half of the sixteenth century, a fantastical overburdening epic poem, is "light-hearted, superficial, ironic, playful, popular", while Tasso's *Gerusalemme liberata*, an epic poem on the First Crusade, written half a century later in 1581 in the context of the Counter Reformation, is "wordy, elevated, melancholy, psychologically penetrating".[8] Galileo, who had no sense for the psychological depths of Tasso's characters, clearly sympathized with Ariosto. Heilbron sees in this preference a reflection of his way of doing science and his predilection for mathematics:

> As in literature, so in physics Galileo dealt more comfortably with the accidents than with the essences of things . . . Galileo's ducking of causal connections in physics and his discomfort with depth of character in literature, and his reliance on mathematics in offense and defense, had the same psychological roots. Initially the ducking, the discomfort, and the reliance protected him from risk-taking; but in time they blinded him to the risks he ran.[9]

Failing to understand and appreciate the complex psychological motivations of Tasso's figures, Galileo charged Tasso with "poverty of invention", while he regarded Ariosto's work as "magnificent, rich, and marvelous". In Ariosto, as in a cabinet of curiosities, Galileo found an endless number of marvelous objects, ranging from "a royal gallery with a hundred ancient statues by the most celebrated sculptors" to "a great many vases of crystals, agate, lapis lazuli and other gems". Furthermore, there were all kinds of fantastical events happening on the earth and in the heavens, ranging from giant sea monsters to a voyage to the moon. Heilbron infers that "Galileo prized inventiveness, when it produced great and beautiful things".[10] Ariosto's fantastical worlds may well have inspired Galileo to further push his astronomical and scientific research into new dimensions.

Galileo's "aversion to complexity in character" may also explain his focus on reason, experience, and facts and his "compulsive attention to linguistic details".[11] Galileo was eager to point out errors, to dismiss pedantry, and to criticize wrong belief in authority. The issue of verisimilitude ranked high in contemporary literary discussions, and Galileo joined these debates. Galileo preferred Ariosto because he was closer to truth in the end. For Galileo, Tasso's errors and inconsistencies stuck out, as he did not respect the laws of mathematics and physics and failed to provide a plausible geography for his plot. Ariosto, on the contrary, made no false pretense of truth; he simply treated the marvelous as if it was true.

The issue of verisimilitude also played a role in Galileo's discussion about Dante's *Divina Comedia*. Dante's *Divina Comedia* was the subject of a literary debate in the 1580s that examined this work in light of the criteria for poetics laid out by the authority of Aristotle and Horace. At stake were the work's "unity of action, coherence, and plausibility", as well as its capacity to offer "pleasure and utility with appropriate decorum, imitation, and invention".[12] The Accademia Fiorentina provided a forum for debates on what poetry was about, with one side arguing for credible imitation of matters scientific, philosophical, and theological, as long as it was done in a poetic way, while the other side condemned such an imitation as sacrificing the essence of poetry. The Accademia chose Galileo to intervene and to analyze the geography and geometry of the inferno.

Reading Ariosto not only filled Galileo's imagination with wonders and marvelous things, but also shaped his writing. Galileo aimed to imitate Ariosto's "clean, crisp, limpid, precise, assured, ironical, natural, direct style".[13] Clarity and plausibility in writing, acquired thanks to years of literary studies, helped Galileo put his surprising scientific results into convincing form. Ariosto's imaginary worlds appeared to be real, just as Galileo's revolutionary geometry of the heavens and physics of the earth would appear to be real in their textual forms. Heilbron describes nicely the "slide from the hypothetical and probable to the true and necessary", from literary technique to scientific knowledge (which may be even more imaginative and ultimately supersede):

> Years of reading the poets and experimenting with literary forms were not mere sidebars—they enabled Galileo to write clearly and plausibly about the most implausible things . . . The realistic treatment of the marvelous in Ariosto's style became a frequent and powerful literary technique with Galileo. His appreciation of tall tales told realistically would help him to slide easily from the hypothetical and probable to the true and necessary, as in the eventual rendition of the Copernican system as something akin to revealed truth.[14]

Indeed, Galileo did not just succeed in communicating his discoveries; he communicated them brilliantly and is counted today among the best Italian writers, together with Machiavelli and Manzoni. The first of his three masterpieces, the *Sidereus Nuncius* (1610), was written in Latin, which he abandoned for his two other books, *Dialogue Concerning the Two Chief World Systems* (1632) and *Discourses and Mathematical Demonstrations Relating to Two New Sciences* (1638), written in Italian.

For Galileo, language and literature thus served as an inspiration for both investigation into nature and elegant and forceful communication. Historically, Galileo made a decisive step when he searched for marvels outside texts, and in nature itself, and he possessed the necessary linguistic skills to make this extraordinary exploration plausible to a stunned audience. At the same time, language was not enough: inquiries into nature also required materials and instruments, on the one hand, and mathematics and geometry on the other. However, Galileo did not see the radical break here that we see today in hindsight. He rather was in continuity with a tradition that saw *logos* as both—word and calculation—and presented mathematics as just another, yet superior, language, when he wrote in *The Assayer*:

> Philosophy is written in this grand book—I mean the universe—which stands continually open to our gaze, but it cannot be understood unless one first learns to comprehend the language and interpret the characters in which it is written. It is written in the language of mathematics, and its characters are triangles, circles, and other geometrical figures, without which it is humanly impossible to understand a single word of it; without these, one is wondering in a dark labyrinth.[15]

Ultimately, Galileo's desire for clarity and unmistakable truth pushed him away from ordinary language, which may lead into a "dark labyrinth". Still, it is the language and literature of his time, which had contributed to pushing Galileo beyond the borders of existing knowledge. Here, Galileo had creatively constructed a new role at the intersection of culture and investigation of nature, inaugurating a new way of doing natural philosophy.

2. HEINRICH KAYSER AS A PHILOLOGIST OF PHYSICS

Whereas Galileo's science emerged from a classical culture, Heinrich Kayser's late nineteenth-century experimental physics seems at first sight completely detached from classical culture. Kayser represents the persona of a German *Bildungsbürger* par excellence in that he self-consciously cultivated classical and linguistic knowledge in his leisure time while earning a living as a physicist in the *Kaiserreich*. Whereas for Galileo there was no scientific alternative to the classical literary and cultural world of his time, Kayser happily worked in a world ruled by science, instruments, objectivity, and reason, which, however, in his view needed to be complemented by the beauty of literature and language to fulfill the ideal of wholeness of personal development.

Heinrich Kayser studied physics at the University of Strasbourg in the 1880s before being appointed as professor of physics at the Technical University of Hanover in 1885 and later occupying the prestigious chair of physics at the University of Bonn. He worked in the domain of spectroscopy, recording and photographing spectra of elements and studying possible regularities within or among spectra. A worldwide specialist in this domain, he remained exclusively focused on this line of work until his retirement after the First World War. Kayser's autobiography, *Erinnerungen aus meinem Leben*, written during the 1930s, provides insights into the lifeworld of a *Kaiserreich* professor of physics. Kayser had a classical education and was highly interested in the study of languages, even briefly flirting with the idea of becoming an Egyptologist while a student at the University of Strasbourg. In line with a European fascination since the Renaissance for Egyptian culture, Kayser even traveled to Egypt, and deciphered and translated Egyptian texts. Nor was Kayser alone in this as a scientist. Leading nineteenth-century scientists such as François Arago, Charles Fourier, Jean-Baptiste Biot, and Hermann von Helmholtz were in close contact with Egyptologists. Physicists, Egyptologists, and philologists met in professorial circles. Kayser's autobiography details his multiple activities in such circles; his travels to countries like Italy, Greece, Turkey, Egypt, England, and the United States; his effort to learn foreign languages, such as Italian, English, Russian, French, Persian, and Sanskrit; and his strong interest in Japanese culture.[16]

Kayser clearly separated his experimental physics from his interests in languages in his autobiography, although his understanding of language nevertheless informed his way of doing physics, ending in something that may be called a "philology of physics".[17] Kayser's empirical work in experimental spectroscopy treated spectra like texts: he recorded, read, and edited spectra in a highly systematic manner, just like a philologist did. The complex fragments of earlier Egyptian languages with their visual and mystical signs required painstaking work of accuracy, just like the thousands of lines in spectra. Then the regularities within and between the lines (or texts) had to be identified. The Egyptian hieroglyphs contained a secret message, and the spectral lines equally contained not yet decipherable information about atoms or molecules. As Kayser firmly believed that the natural world followed laws, he therefore thought the regularities among spectra lines contained some meaning. However, Kayser's philological approach in physics could not deliver any meaning, as it remained restricted to reading phenomena only on the surface without providing meaning within a theoretical framework. In the end Kayser's physics succeeded only on the experimental level, providing first-class spectra by using refined techniques and skills, but this work accumulated only data, and texts on spectra, which he compiled in a most exhaustive way in the eight-volume *Handbook of Spectroscopy*.[18] Decipherment did not come by way of a more detailed analysis of spectra, but by way of an atomic model and theoretical synthesis, provided by the Danish physicist Niels Bohr in 1913. Kayser's philological approach to physics failed to come up with meaning, as he failed to realize that mere numbers would not take him directly to meaning in the physical world. A classical philologist, in contrast to a philologist of physics, had the advantage of being able to go back and forth between familiar und unfamiliar languages. Kayser succeeded in registering numerical analogies in the field of spectroscopy, but the causes of these mysterious regularities remained beyond his grasp. The German physicist Arnold Sommerfeld stated in the preface of his famous book *Atombau und Spektrallinien* of 1919:

> What we are nowadays hearing of the language of spectra is a true 'music of the spheres' in order and harmony that becomes ever more perfect in spite of the manifold variety. The theory of spectral lines will bear the name of Bohr for all time. But yet another name will be permanently associated with it, that of Planck. All integral laws of spectral lines and of atomic theory spring originally from the quantum theory. It is the mysterious organon on which Nature plays her music of the spectra, and according to the rhythm of which she regulates the structure of the atoms and nuclei.[19]

In Sommerfeld's phrase the reference to language is merely metaphorical, and music fits as a much more suited analogy to represent physical regularities, now explained with Planck's quantum theory and Bohr's atomic model. For Sommerfeld, the laws of nature were best presented to students of science by writing about them in terms of harmony and rhythm of the atoms.

3. WERNER HEISENBERG AND HIS FIXATION ON LANGUAGE

The third case study deals with one of the architects of twentieth-century theoretical physics, Werner Heisenberg. Like Kayser, Heisenberg was a German *Bildungsbürger* who knew his Goethe. Unlike Kayser, Heisenberg struggled throughout his life to make sense of the mathematical formalism of quantum mechanics, a struggle that led him to reflect repeatedly on the limits of language for putting modern physics into words. Heisenberg's father was a philologist and professor of Byzantine philology at the University of Munich. As a pupil, the young Heisenberg enjoyed an excellent humanistic (classical) education, studying thoroughly Greek and Latin, learning to play the piano at quite a high level, and doing rather little in the natural sciences. As a university student, however, Heisenberg eagerly turned to theoretical and mathematical physics, receiving his doctorate under Arnold Sommerfeld (quoted earlier) in 1923 and becoming a full professor of theoretical physics at the University of Leipzig in 1927 at the young age of twenty-six.[20]

As a gifted young scientist, Heisenberg pushed himself to the limits, retreating from the world and immersing himself for weeks or months exclusively in numbers, formulas, and abstract thinking. These studies resulted in the development of a new physical interpretation of atoms, quantum mechanics, which was based solely on mathematical formalism. However, with quantum mechanics came a conundrum, already identified by the Danish physicist Niels Bohr: the "failure" of both everyday language and the language of classical physics to describe physical events on the level of the atom. Faced with a tension between subjective experience of the everyday world and objective scientific formalism, Heisenberg could have resigned himself to a position that the natural scientist, unlike the poet, "never gets at the *essence* of things", as the young Heisenberg wrote to a friend in 1924.[21] He did not. Cathryn Carson stresses that he chose, just like his close colleague Niels Bohr, "to keep on the trail of deep meaning in his science".[22] It is therefore not surprising that during the 1930s "language (*die Sprache*) grew into a dominating theme in his thoughts about science".[23] In an ongoing personal quest, Heisenberg struggled to come to terms with the ultimate meaning of his discoveries, which would integrate into the familiar classical culture of the *Bildungsbürger*.[24]

Heisenberg played a prominent role in the public cultural sphere of post-war Germany, when physicists and philosophers enjoyed high prestige and received numerous invitations for public lectures. But for Heisenberg there was not only the outside stimulus for communicating his new-won science; he also felt deep from within this necessity to put his ideas into some form, whether literary or philosophical. For example, when rather isolated during the Second World War, he wrote a manuscript *Ordnung der Wirklichkeit* (*The Ordering of Reality*), which he distributed only among friends.[25] The manuscript was no less than an attempt to build upon Goethe's ideas about levels of reality as formulated in the *Farbenlehre*, ranging from classical physics over quantum physics and consciousness to the spheres of the symbolic and creative, with each sphere accompanied by a language that "transformed from the static, precise, but content-poor language of mathematical physics to the dynamic and fertile languages of the symbolic and the creative". Carson sees here the typical "*bildungsbürgerliche* web of subjectivity and objectification", where "science was knit together with religious and artistic experience, love and alienation".[26] Heisenberg acknowledged that science could not give access to the meaning of life, but at least, he hoped, it may point to the "hidden harmonies of the world".[27] Beyond these philosophical-religious reflections, language figured prominently in this text. In line with Wilhelm Humboldt, Heisenberg attributed to language a formative role when he wrote: "Every linguistic formulation of reality does more than just grasp, but also molds and idealizes".[28] Language made thinking possible, as Carson underlines:

> Without language, no thinking; without language, no philosophy manuscript. But what made thinking (and philosophy manuscripts) possible, was not necessary static scientific precision. Rather it was the 'ambiguity

> and indeterminacy [*Mehrdeutigkeit* und *Unbestimmtheit*]' inherent in language's fertile richness. Meaning comes also from context, literary construction, and the resonating copresence of unspoken ideas . . . [Heisenberg] was calling on a whole complexly fluid set of semideterminate meanings in which metaphor and allusion played critical roles.[29]

But language for Heisenberg was not only about subjective expression, it was also about communication: "Language evoked the communality of science, the fact that it was something larger than individual persuasion—intersubjectivity, in short".[30] Language served for Heisenberg as a vehicle mediating between two worlds, the subjective and objective, the inside and outside, the physicist and the public. Heisenberg emphasized the importance of communication over and over in his more public and philosophical texts, particularly the exciting informal discussions among physicists in the late 1920s. Talking meant making sense of mathematical formalism, making sense of the world. For Heisenberg, meaning is produced in the act of speaking, in intersubjectivity (not in experience, as positivists contended).

The theme of communication and intersubjectivity was also central in Heisenberg's most engaging public reflections about language, and quantum mechanics, which grew out of the public Gifford Lectures of 1955/1956 and ultimately appeared in 1960. In quantum mechanics, Heisenberg explained, common language fails to express the abstract mathematical language of symbols. However, Heisenberg stressed, "We have to talk about them [atoms and elementary particles], because otherwise we could not understand our experiments".[31] Heisenberg pointed here to the paradox of quantum mechanics that we have to describe physical experiments in classical concepts; in reference to Bohr's interpretation, he wrote that "we have to use different modes of representation in atomic physics that are mutually exclusive but complementary, so that only through playing with different images will we obtain an adequate representation of an event".[32] The reference to playing (*Spielen*) probably came from Wittgenstein's notion of "game". Like the older Wittgenstein, Heisenberg discarded a positivistic view of language and warned that reducing language to logic would end up in a too great simplification and impoverishment of the actual possibilities of language, of the actual use of language, as in literature. He also pointed to the fact that children learn language and meaning by the *use* of words and that language provided a possibility to create a "web of semi-conscious representations", superior to logical deduction:

> The words are, so to speak, tools to orient ourselves in surroundings, to get in contact with them . . . It is possible that this web of semi-conscious representations, which can be evoked by language, reflects better the meaning of what is intended to be said than a sharp logical conclusion.[33]

Heisenberg finished his article with a discussion of the possibility of a new language, as suggested by von Weizsäcker's quantum logic, which aimed to replace classical Aristotelian logic:

> We are therefore forced to learn a new language, which in many places is foreign to common language. However, a new language also means a new way of thinking, and hence natural science puts the claim forcefully on the table, which in our time is posed in so many realms of life.[34]

Heisenberg's reflections on language remained fragments. Kristian Camilleri sees an "unresolved tension" in Heisenberg's reflections on language, as Heisenberg attributed to language a world-disclosing function, as well as a designative one.[35] However, it is also possible, like Carson has done, to attribute this ambivalence to Heisenberg's style of writing in German language, which left many things 'between the lines', making any "straight-ahead reading" difficult for the reader, and which explains the "extraordinary divergence of interpretations that have been put on his texts".[36] Whatever it may be, Heisenberg's persistent reflections on language were the necessary complement to his revolutionary work in physics, as only language provided the key to attributing meaning and thus making sense of the world.

4. CONCLUSION

These case studies lead to more questions than answers. I would like to raise three issues for possible future investigation. First, one of the puzzles of the emergence of modern science is how classical education, with language and literature at its core, helped give birth to something so strange and unlikely as modern physics and science. The cases of Galileo and Heisenberg raise the question of to what extent a firm anchorage in classical culture was a necessary condition for revolutionary work, giving as it did the confidence, assurance, and means to identify and explore unknown continents. Galileo knew intimately the rules of debates and writing, of persuasion and beauty. This may explain both his initial reluctance to publicly advocate his scientific results and his later persistence in them, once he had enough revolutionary material that was difficult to refute off-hand. The young Heisenberg, endowed with prodigious skills in mathematics, set sail for new continents from the safe harbor of a solid classical education. The German ideal of *Bildung* encouraged him to find his own ways for personal self-fulfillment, and Heisenberg pushed it to the extremes by immersing himself exclusively in mathematical formalism. In a second step, however, he then felt the urgent need to tell fellow scientists and the public about this new continent, leading him to reflections on language that aimed to integrate his revolutionary results into the culture he originally came from. Galileo and Heisenberg both clearly saw that their work was new and provocative, but they both worked hard to bring the new into the culture of their time by relentless communication and by addressing large audiences. All three physicists published in literary genres, ranging from dialogues to essays to autobiographies to explain their work, life, and world view to larger audiences.

Second, the uses of language are manifold and specific to each historical situation. Language changes over time and place, and so does science. But their developments are not in step: language and the study of language may inspire scientific research at one point and hamper it at another. Inversely, new scientific results may change language and inspire new studies of language and literature. These links are so intricate and twisted that they can probably be traced only in highly focused studies. To complicate things even more, philology, linguistics, literature, and philosophy of language, which interact with language and science in various ways, have thus far attracted little systematic attention.

Third, scientists' attention to language (or to certain aspects of it), and their belief in its force and pertinence, shifts over time and depends on the larger cultural context. It remains an open question whether Galileo's and Heisenberg's strong desire to give something back to the society at large is characteristic of European culture and its firm anchorage in language and literature. Scientists in other cultures may not feel this urgent need to put their scientific works into lectures, articles, or literary texts. Unlike the three physicists discussed here, they may have no in-depth training in languages and literature and not be at all in contact with linguists, philologists, or literary persons on a personal level. Thus cultural conditions may encourage or discourage scientists to explain their work in words. In contrast with the European physicists Heisenberg and Bohr, American physicists of the period felt much less of a need to put quantum physics into words. Thus, searching for the world-disclosing capacities of language (as a poet would do) and attributing meaning to scientific results may only make sense for scientists who already have an audience for these kinds of reflections, as was the case, for example, in Germany for Heisenberg.

It may also be that attention to and emphasis on language may shift over the life of a scientist. Richard Rorty argued that, in philosophy, Ludwig Wittgenstein and Martin Heidegger switched their view on language and words as tools:

> The older Heidegger retreated from sentences and discourse to single words—words which had to be abandoned as soon as they ceased to be hints (*Winke*) and became signs (*Zeichen*), as soon as they entered into relations with other words and thus became tools for accomplishing purposes. The younger—unpragmatical, mystical—Wittgenstein had wanted sentences to be pictures rather than merely tools. By contrast, the pragmatical young Heidegger, the philosopher of inescapable relationality (*Bezüglichkeit*), had been content to let them be tools. But the older, more pragmatical Wittgenstein became content to think of them

> as tools, about the same time that the older Heidegger, decided his early pragmatism had been a premature surrender to "reason [which], glorified for centuries, is the most stiff-necked adversary of thought".[37]

In the realm of scientific research, shifts in emphasis on language happen in parallel with changing cultural, aesthetic, and philosophical fashions and constraints. These subtle shifts may guide scientists to new topics of research, or they may be helpful for reflections on the meaning of new scientific results. Science and language share two crucial features: they are permanently in flux and endlessly malleable for all kinds of creative purposes.

NOTES

1. For a recent book on the history of languages of science, see Michael D. Gordin, *Scientific Babel: How Science Was Done Before and After Global English* (Chicago: Chicago University Press, 2015).
2. Edmund Husserl, *Die Krisis der europäischen Wissenschaften und die transzendentale Phänomenologie* (The Hague: Martinus Nijhoff, 1976), 2. ed. (Husserliana, vol. VI).
3. Cathryn Carson, *Heisenberg in the Atomic Age. Science and the Public Sphere* (Cambridge: Cambridge University Press, 2010); John L. Heilbron, *Galileo* (Oxford: Oxford University Press, 2010).
4. Matthias Dörries, "Heinrich Kayser as Philologist of Physics," *Historical Studies in the Physical and Biological Sciences* 26, no. 1 (1995), 1–33.
5. Ludwig Wittgenstein, *Philosophical Investigations* (London: Wiley-Blackwell, 2009).
6. Heilbron, *Galileo*, 1.
7. Heilbron, *Galileo*, v.
8. Heilbron, *Galileo*, 18.
9. Heilbron, *Galileo*, 22.
10. Heilbron, *Galileo*, 19.
11. Heilbron, *Galileo*, 21.
12. Aristotle, quoted in Heilbron, *Galileo*, 14.
13. Heilbron, *Galileo*, 22.
14. Heilbron, *Galileo*, 21.
15. Galileo in: Stillman Drake and C.D. Malley, *The Controversy on the Comets of 1618* (Philadelphia: University of Pennsylvania Press, 1960), pp. 183–184; Galileo Galilei, "Il saggiatore," in *Opere*, ed. Antonio Favaro, 20 vols. (Florence: Barbera, 1890–1909), 6, p. 232.
16. Heinrich Kayser, *Erinnerungen aus meinem Leben*, eds. Matthias Dörries and Klaus Hentschel (Munich: Institut für Geschichte der Naturwissenschaften, 1996), 323, 225–6, 279–80.
17. Dörries, "Heinrich Kayser as Philologist of Physics," 2.
18. Heinrich Kayser, *Handbuch der Spektroskopie*, 8 vols. (Leipzig, 1900–1934).
19. Arnold Sommerfeld, *Atomic Structure and Spectral Lines* (London: Methuen, 1923), viii. "Was wir heutzutage aus der Sprache der Spektren heraus hören, ist eine wirkliche Sphärenmusik des Atoms, ein Zusammenklingen ganzzahliger Verhältnisse, eine bei aller Mannigfaltigkeit zunehmende Ordnung und Harmonie. Für alle Zeiten wird die Theorie der Spektrallinien den Namen Bohrs tragen. Aber noch ein anderer Name wird dauernd mit ihr verknüpft

sein, der Name Plancks. Alle ganzzahligen Gesetze der Spektrallinien und der Atomistik fließen letzten Endes aus der Quantentheorie. Sie ist das geheimnisvolle Organon, auf dem die Natur die Spektralmusik spielt und nach dessen Rhythmus sie den Bau der Atome und der Kerne regelt". Arnold Sommerfeld, *Atombau und Spektrallinien* (Braunschweig: Vieweg, 1919), viii.

20. David Cassidy, *Uncertainty: The Life and Science of Werner Heisenberg* (New York: Freeman, 1992).
21. Carson, *Heisenberg*, 44.
22. Carson, *Heisenberg*, 44.
23. Carson, *Heisenberg*, 87.
24. For an excellent introduction into the notion of Bildung und Bildungsbürger, see chapter 3 of Carson, *Heisenberg*.
25. Werner Heisenberg, *Die Ordnung der Wirklichkeit* (Munich: Pieper, 1989).
26. Carson, *Heisenberg*, 55.
27. Heisenberg, quoted in Carson, *Heisenberg*, 57.
28. Heisenberg in *Ordnung*, 146, quoted in Carson, *Heisenberg*, 45.
29. Carson, *Heisenberg*, 57 and 87–88.
30. Carson, *Heisenberg*, 88.
31. Werner Heisenberg, "Sprache und Wirklichkeit in der modernen Physik," in *Gestalt und Gedanke* (Band 6. Jahrbuch der Bayerischen Akademie der schönen Künste) (Munich: Oldenbourg, 1960), 32–62, on 49.
32. Heisenberg, "Sprache und Wirklichkeit in der modernen Physik," 51: "daß wir in der Atomphysik verschiedene Beschreibungsweisen verwenden müssen, die sich zwar gegenseitig ausschließen, aber doch auch ergänzen, so daß erst durch das Spielen mit den verschiedenen Bildern schließlich eine angemessene Beschreibung des Vorgangs erreicht wird".
33. Heisenberg, "Sprache und Wirklichkeit in der modernen Physik," 34, 37: "Die Wörter sind gewissermaßen Werkzeuge, um sich in der Umgebung zurechtzufinden, um mit ihr in Kontakt zu treten . . . Unter Umständen kann gerade dieses Gewebe von nur halbbewußten Vorstellungen, das durch die Sprache hervorgerufen werden kann, den Sinn dessen, was ausgesprochen werden soll, besser wiedergeben als ein scharfes logisches Schlußverfahren".
34. Heisenberg, "Sprache und Wirklichkeit in der modernen Physik," 62: "Wir sind daher gezwungen, eine neue Sprache zu lernen, die der gewöhnlichen Sprache an vielen Stellen sehr fremd ist. Eine neue Sprache bedeutet aber auch eine neue Art zu denken, und damit wird in der Naturwissenschaft in aller Schärfe die Forderung erhoben, die sich in unserer Zeit so sichtbar in so vielen Bereichen des Lebens stellt".
35. "Critical here was Heisenberg's commitment to the world-disclosing function of language, though Heisenberg retained something of a realist perspective in recognising that language has a designative function, that is to say, language refers to a world which exists independently of how we conceive of it". Kristian Camilleri, *Heisenberg and the Interpretation of Quantum Mechanics: The Physicist as Philosopher* (Cambridge: Cambridge University Press, 2009), 152–153.
36. Carson, *Heisenberg*, 10.
37. Richard Rorty, "Wittgestein, Heidegger, and the Reification of Language," in *Essays on Heidegger and Others*, ed. Richard Rorty (Cambridge: Cambridge University Press, 1991), 50–65, on 52.

Part 1

Language, Rhetoric, and History

3 How Language Became a Tool

The Reconceptualisation of Language and the Empirical Turn in Seventeenth-Century Britain

Miles MacLeod

The central theme of this volume is that language has operated for science over the course of its history and in many diverse contexts as a tool to be applied to purposes both epistemic but also rhetorical and political. In this paper I want to show that the conditions for the treatment of language as a tool in many of these respects actually required a shift in the understanding of language around the seventeenth century. This shift reflects the modern understanding of language in the minds of scientists and researchers. It certainly affected other seventeenth- and eighteenth-century contexts directly (see for instance Herrnstadt and Schlicht, Sumillera, this volume) for which the work of John Locke and other empiricists was essential to rationalizing the specific uses to which language is put. The understanding of language I am referring to is expressed simply enough as the notion that there is a firm distinction between the word and the thing represented. The empirical turn that characterises the enlightenment in British natural philosophy in the seventeenth and eighteenth centuries relied upon just such a reconceptualisation of language. This reconceptualisation reduced language to a system of arbitrary or conventional signs and symbols for purposes of representation and communication. It had many functions for the natural philosophers involved: ontological, epistemic, and rhetorical. While this way of thinking about language today might seem an obvious way of thinking about language in science, to those at the time it was no means uncontested and required a novel reflection and consciousness of language by philosophers like Bacon, the Royal Society members, and particularly Locke. It led ultimately to the need for the development of a suitable philosophy of language for natural philosophy.

In this paper I will explore how this shift came about, the nature of it, and what motivated it for the developers of the natural historical tradition from Bacon, Boyle, through to its most complete articulation with John Locke. I hope to show in the process that our standard picture of the empirical turn in Britain as the result of an empirical revolution cannot be fully understood without the linguistic revolution that made it possible. The natural history before this revolution of the likes of Gesner and Versalius was built on a different understanding of language distant from the neutral tool-like conception the British empiricists sought. This understanding treated moral meaning as the fundamental essence of nature and in this respect saw no important distinction between words and natural things. At the same time the speculative natural philosophy of the Schools elevated logic and grammar to the pinnacle as sources of information about the world. Language thus had to be reconceived as a tool in order to shake it free of all these associations so that the broader philosophical transformations of empiricism could function. Part 1 will concern the intellectual context against which this reconceptualisation of language occurred, framed on the one hand by scholasticism and on the other by the "Emblematic World View". In part 2, I will discuss the developments by which language in the context of natural historical study came to be first "desymbolified" and then philosophically reformed.[1]

1. A REMARKABLE SHIFT

Something important happens to the way language is perceived by natural historians and experimentalists in Britain in the 1600s. For example in 1603 Alexander Top in his work *Oliue Leafe*, a natural history of words, pronounced in respect of Hebrew, the language closest to that of God, that "no word in that tongue ever fayleth of some proper or distinct meaning; as well to instruct the diligent Schollar in some natural reason of things which is the truth of all Language in deed".[2] Edward Topsell in his *A Historie of Four-Footed Beastes* expressed that "a history of beasts should be preferred to an historical chronicle, because that Chronicle which was made by God himselfe, every living being a word, every kind being a sentence, and al of them together a large history, containing admirable knowledge & learning".[3] By the end of the century, however, we have this from John Locke, himself an avid natural historian, in his *Essay Concerning Human Understanding*:

> Thus we may conceive how *Words*, which were by Nature so well adapted to the purpose, came to be made use of by Men as *the Signs of* their ideas; not by any natural connexion that there is between particulate articulate sounds and certain *Ideas*, for then there would be but one Language amongst all men; but by a voluntary Imposition, whereby such a Word is made arbitrarily the mark of such an *Idea*.[4]

One thing that leaps out from this quotation is the fact that the meaning of words, formerly fixed by nature and God, has changed to something arbitrary, imposed by man according to his purposes. The dramatic change instituted the idea of language as a tool in the context of experimentalism and natural history. Why and how such changes became possible, plausible, and necessary to some British philosophers requires insight into the conceptions of language current at the time.

1.1 Renaissance Natural History and the "Emblematic World View"

The emblematic world view, documented by Ashworth,[5] but also famously by Jacob,[6] was the dominant tradition of Renaissance natural historical scholarship from about 1550 to the 1660s and shared a similar perspective on language to the Paracelsian and hermetic tradition. As Ashworth is keen to point out, scholarship has underplayed the part of reformations of natural history in the scientific revolution and enlightenment, preferring to focus on developments in physics, chemistry, and experimentation. This seems doubly true when we consider in the British context that the Royal Society and particularly Locke himself took natural history as central and proper natural philosophical work. Those who do cite natural history in their analyses of the origins of the scientific revolution tend to find continuity with natural historical trends between the sixteenth and seventeenth centuries by singling out the work of Belon (1517–1564) and Guillaume (1538–1616) and their remarkable dissections and anatomical studies.

But the dominant natural historical tradition at the time was rather different from these. It involves figures like Gesner (1516–1565) and Aldrovandi (1522–1605). The natural histories (principally of animals) composed by these authors are not like any that exist today under the heading of "natural history", and certainly nothing like the conception of natural history we find in writers like Boyle and Locke. What counts as important for the description of the beasts of the world are not simply behavioural features, and the anatomical and physiological parts that distinguish animals but a font of information about the relation of a creature to the human universe of morality, history, and of course God's divine order. Hence Gesner's discussion of various animals in the volumes of the *Historiae Animalium* (1551–1558) is not driven by their anatomical features, but rather by their cultural history, covering the peculiar characteristics and habits they were believed to possess (a peacock, for instance, is afraid of its feet; its flesh does not decay after death). These histories, for example, documented the meaning of expressions invoking the animal; its part in fables, legends, and mythology; its use in recipes and medicines. All such statements were supported by ancient authorities like Pliny and Aristotle.

One should not interpret these analyses as just cultural histories, as though "natural histories" or "nature" itself could be rendered as something distinct (as we today think). For these authors these were "natural histories". Their view of nature derived from a neo-platonic tradition of seeing animals as units of meaning (symbols, metaphors, and analogies) within God's language of moral order. As something of a research paradigm, these natural historians pursued the collection of such information from Gesner's time on, expanding and categorising it. By the time of Aldrovandi what was an eight-page treatise on the peacock in Gesner's work is thirty-one pages, with a plethora of categories for its elucidation:

moralia, symbola, genus, synonyma, differentia, incubatus, vox, locus, proverbial, mystica, fabulosa, historica.[7] As noted, Ashworth refers to the perspective underlying this endeavour as the emblematic world view, a "belief that every kind of thing in the cosmos has myriad hidden meanings and that knowledge consists of an attempt to comprehend as many of these as possible".[8] Such information was to be received for humanists through cultural sources like the representations of animals in hieroglyphs, fables, antique coins, classical mythology, adages and epigrams, and emblems themselves, for which Alciati's *Emblemata* (1531) collection was the foremost source.

This close association between objects of nature (like animals) and the meaning of words betokened a different perception of language than is familiar today, at least in the realm of modern science. It found no important distinction between the word, its part within literary culture and traditions, and the object of nature itself. Nature was a moral language written by God to instruct humankind. Hence the sign and what it represented were to be treated in this sense as ontologically equivalent. Both were objects of nature, and both were objects of language. We can thus tease out the following qualities to the understanding of language in this emblematic tradition. First, words were themselves natural in the sense that no significant ontological distinction was to be made in this regard between the reality of language and the reality of the world. Language was rather part of the natural world and had devolved with things from creation. Second, the correspondence between things and the meanings of words was intrinsic, to be treated on the same level as a relation between natural things themselves. God had ascribed this relation in Creation. Third, words were interchangeable with their objects. Objects in the world were themselves signs having thoroughly linguistic character. This last was referred to by Elsky as the analogical structure of reality.[9]

1.2 The End of Emblematic Tradition

However, by the beginning of the seventeenth century, certain alternative perspectives were taking shape in natural historical study, at least in Britain, which reflected a dramatically different vision of the world and particularly of the importance of language. Ashworth claims three factors were responsible for this in Britain. First, the influx of creatures and plants from the new world that simply had no cultural or literary history and could not be studied meaningfully in this respect; second, the subjection to test by Thomas Browne (1605–1682) of many of the beliefs that abounded in these natural histories, with negative results (peacock flesh does decay); third, the rise of Baconism. These were all part of a reconceptualisation of nature as something distinct from and independent of the human moral sphere. As Ashworth reports, historical linguists refer to this as a process of "decontextualisation, disenchantment and desymbolification".[10] Rawley, for instance in his introduction to Bacon's own natural history *Sylva Sylvarum* (1627), makes the following comments on those prior views:

> For those Natural Histories which are Extant, being gathered for Delight and Use, are full of pleasant Descriptions and Pictures . . . But contrariwise, the Scope which his Lordship intendeth is to write such a Naturall History, as may be fundamental to the Erecting and Building of a true philosophy.[11]

By the time of Joannes Jonston's *Historia Naturalis* (1650), a text of some influence with the Royal Society and considered the fundamental turning point in Foucault's own historical accounts,[12] we find none of the concern with emblems, hieroglyphs, or proverbs, but rather something we might recognise ourselves as a natural history, namely, an encyclopaedic concern with identifying and classifying the variety of species and giving accurate visual representations that show morphological differences and similarities. "Living bodies were scraped clean so to speak. They shook off their crust of analogies, resemblances and signs, to appear in all the nakedness of their true outer shape".[13] Very quickly the sixteenth-century style of natural history disappeared in Britain, to be replaced by those of Royal Society participants like Hooke and Boyle.

1.3 Scholasticism and Words

Scholasticism also is an important part of the background context against which the reconceptualization of language took place. Indeed, in the rather cloistered but still elite Schools the consideration of language and its role was much different for those of the old natural historians, but in many ways was just as problematic to the new natural historians and experimental philosophers. Aristotle himself bequeathed something like a referential or correspondence theory of language, which made a distinction between the sign and its referent, but with his own metaphysical presumptions added in. The strong commitment of the schoolmen was to the metaphysical importance of logic, particularly syllogistic logic, which served to put language at least in this form at the top of the ontological tree, from which all else flowed. The modists of the thirteenth and fourteenth centuries had asserted, following Aristotle, that the grammatical structure of natural language reflected the structure of reality and that, through an understanding of the categories of language, an understanding of the categories of reality would follow. Scholasticism, particularly with its emphasis on the primacy and importance of logic in understanding the natural world, is thus an important background element for understanding the reformation of language from which British empiricism arose. Like the "emblematic world view", the reconceptualisation of language challenged both its ontological and epistemological commitments.

2. THE RECONCEPTUALISATION OF LANGUAGE

The claim of this chapter is that the shift in natural history and the rise of experimentalism in Britain, in opposition to both scholasticism but also the humanist traditions of natural history (including hermeticism), necessitated an active reconceptualisation of the role and importance of language, which cleared the intellectual space for the empirical and ontological shifts historians and philosophers have mostly focused on. This reconceptualisation rendered both scholasticism and the humanist traditions untenable ways of studying the natural world. The classical perspective on this period in the history of British science, particularly that surrounding the Royal Society, evokes typically a methodological (epistemological) and ontological revolution, denoting a shift to the experimental or "factual" given new literary and material technologies, as well as a shift to mechanistic explanation, in which material reality is composed of fundamental units (or corpuscles). Those more interested in contextual explanations look to the institutional and communal organisations of the Royal Society and its followers, as well as the dynamics and freedoms of restoration England. However, facilitating these developments was a sustained effort to reconceptualise language explicitly as a tool. The British empiricists were deeply conscious of the role of language in both the misunderstandings of the past and also the way in which natural philosophy needed to be reformed. Locke dedicated the entire third book of his *An Essay Concerning Human Understanding* to the subject of words, a concern with language unprecedented at the time. He treated it as a necessary part of a natural history of human understanding, thus central to our understanding of human knowledge and its acquisition. As such, these changes in the conception of language were not simply the epiphenomena of other reformations and novel practices, but were at the forefront of thinking. In essence they served to free language from the associations of renaissance natural history and scholasticism, which compromised and muddied the kind of project empiricists wanted to run. Further, clarifying language could be made to serve their own rhetorical ends in disputing the value of these other perspectives. In the rest of this paper I look more closely at the form this reconceptualisation took and how it served the ends of the British empiricists in making language safe for natural history and experimentation.

The work of Bacon, Boyle, and other Royal Society participants, but particularly that of Locke, took the form of what would seem to us in modern terms a referential or conventional view of language. It had a historical precedence, as we have noted in Aristotle, and was not original to these philosophers. However, this view got a distinctive development in this British context that was unique to it in terms of its articulation and the functional work it was meant to play. The basis of this viewpoint, starting with Bacon, was a clear separation between signs and objects in the world—or at least ideas. Signs were neither in nature, nor intrinsically related to objects, nor interchangeable with objects. Importantly, language was considered arbitrary and conventional, strictly for the purposes of representing and communication. Meanings and referents were to be fixed by convention. As Locke puts it,

> *Words*, by long and familiar use, as has been said, come to excite in men certain *Ideas* so constantly and readily, that they are apt to suppose a natural connexion between them. But that they *signify* only Men's peculiar *Ideas*, and that *by a perfect arbitrary imposition*, is evident, in that they often fail to excite in others (even that use the same language) the same *Ideas* we take them to be signs of: and every man has so inviolable a liberty, to make Words stand for what *Ideas* he pleases, that no one hath the Power to make others have the same *Ideas* in their minds that he has, when they use the same Words that he does.[14]

This passage is one of the clearest articulations by Locke of his philosophy of language as a tool. The passage is designed to make us rethink our own presumptions about how words operate given a context of muddled ideas about language. Importantly he denies that there are natural connections between words and ideas. Locke stresses the "perfectly arbitrary imposition" of words and also the social-conventional way in which words are affixed to their objects and their meanings transmitted. In Locke's philosophy ideas are directly imposed on us by nature itself and act as the direct referents of words.

Additionally language was properly perceived as fallible like any tool to the extent that the relation between words and things was tenuous and meanings were liable to become lost or confused over time. Locke has two chapters of book III (chapters IX and X) dedicated to the imperfections of words, and their abuse, which derives from the fact that words have "naturally no signification"—their signification is thus subject to human vagaries and error—and operate independently from their ideal referents. They thus can be used in discourses without there being any clear underlying ideas.

How then did this particular tool-like view of language facilitate the British empiricists' own views of the proper study of the natural world at the expense of everyone else's? These tool-like qualities were made to serve at least four principal roles for natural philosophies in the development of the natural historical and experimental philosophies. First, it supported their ontological presumptions while undermining those of competitor groups, such as scholastics and those in the emblematic tradition (like hermeticists). Second, it supported and justified the methodology of natural history and experimentation—i.e., a tool for entrenching their own epistemology, which it seemed well suited to do. Third, it enabled the development of a psychology of language with which to understand how others could be mistaken. Finally, it reinforced the need for minimalism and plain styles given the putative risks of language, which also served to marginalise the competition.

2.1 An Ontological Tool

The chief ontological function of this reconceptualisation of language was to underwrite a conception of nature as distinct from the meanings humans imposed on it. It operated in effect to "de-ontologise" language. There was no longer anything natural about a word through any intrinsic connection between it, the world, and the derived moral order of God in the form of art, literature, and cultural practices. What could be considered natural was distinct from words. Nature was out there and not in the relations of words. Although a word representing the idea of an animal, for instance, might code a moral meaning through literary tropes, this meaning did not transfer to the creature itself through treating the part of a sign in language and place in nature as the same. Words represented what they were assigned to represent. Hence, in this respect, shaking language free of these prior ontological associations cleared the way to an alternative understanding of just what "nature" was. Bacon certainly believed in providential order in the universe, as well as the signatures of the divine upon this order. He did not believe the signs themselves revealed any deeper spiritual meaning. His was rather an ontology of causal relations and hierarchies of classification.

In Locke's *Essay* the ontological significance of the reformulation of language is highly significant. Locke, for instance, relies on the conventional nature of language to project the nominal/real essence distinction, which is central to him defining the limits of natural philosophy and thus its proper methodology. It was common in scholastic circles to hold that species terms referred to the real essences of their collective elements. This putatively inferred a strong semantic connection between substance words and groupings in nature. However, since for Locke words have no immediate connection to things or kinds in the world, so to speak, this type of reference did not exist. It was thus not simply that we did not have the knowledge or observational capacity to access these real essences in the microstructures, but that our terms did not pin to them even so.[15] All our substance categories are rather nominal, and nominal essences are nothing more than abstract general ideas, which prescribe a certain extension of things that fit this idea. Hence, classification is conventional, relying on our ability to abstract, and need not invoke any essentialist scholastic ideas about classification. Words of themselves had no special powers to pick out a deep ontology of nature. As such, a proper understanding of language in combination with his epistemology allowed Locke to constrain the relations between words and reality and undermine the importance of scholastic concepts, which had presumed to affix to real essences whether these were known or unknown. Words just did not do the work presumed of them. They were conventional and thus could only be affixed to the ideas they referenced, which in the case of substances could only be nominal. Real essences were beyond our access and thus not connected to our general terms.

Both Bacon and Locke also challenged the authority of the scholastic syllogism from their revisions of the understanding of language. Bacon enunciated in the *Novum Organum* that Aristotle "enslaved his natural philosophy to his logic, rendering it more or less useless and contentious".[16] For Bacon, syllogism revealed a close reliance between speculative natural philosophy and words which needed to be corrected:

> The syllogism consists of propositions, propositions of words, and words are tokens of notions. Therefore—and this is the heart of the matter—if the notions themselves are muddled and carelessly derived from things, the whole superstructure is shaky.[17]

Locke asks, "Of what use then are *Syllogisms*" and replies, "Their chief and main use is in the Schools, where Men are allowed without Shame to deny the Agreement of *Ideas*, which even to themselves is visible".[18] The idea that the syllogism had a special access to causes could be dispelled by simply reflecting on the role of words and language. Standing in a certain logic relation was not enough; what was important was the initial reference at the outset, and the propositions had to correspond and be clear. For too long scholastic philosophy had elevated the logical structure over the meaning and reference of the terms of that structure.

2.2 An Epistemological Tool

This reconceptualisation of language at the hands of British empiricists also supported their experimental and natural historical methodology, to the exclusion of others. This aspect gets its most extensive clarification from Locke. Within Locke's philosophy (books III and IV of the *Essay*) we find a very clear distinction between language and ideas. Knowledge refers to the ideas we have and the connections we discover between them. It has nothing to do with words themselves. Thus we can assess knowledge claims simply upon the basis of whether they have ideal reference. Ideas themselves are obtained only from sense experience, and they are what science must be built from. It is essential, then, that we realize that words have no direct part in this construction of knowledge.

When the word is a general term it should refer to an abstract general idea, that is, an idea formed from elements of all its particulars. The proper place of language is hence clearly defined: it is instrumental, a container for ideas and relations. The measure of whether a scientific theory is well constructed or not is therefore whether "clear and distinct" sensible ideas underlie the actual words used in the relations supposed. Philosophically, then, the boundaries between experimental science and speculative natural philosophy can be seen distinctly. Locke is enabled to argue from the fact that we lack epistemic access to the world of essences that speculative natural philosophy of the Aristotelian kind can rest on no meaningful basis. The terms and words it employs lack meaning. The preferred mode of science is clearly one that is soundly grounded in ideas, one that accumulates sense data, that is, Baconian methods of natural history and inductive experimentation.[19]

Bacon himself saw part of the task to be the construction of a language (or philosophical grammar) appropriate to natural histories and inductive methodology. This invoked a primary attention to the relations between words and things, "not the analogy of words with one another".[20] Natural languages reflected, for Bacon, as Elsky puts it, "mostly the vulgar misconception of things".[21] Bacon strongly advocated not just the reconceptualisation of language but its redevelopment for these epistemic tasks which inspired movements for universal languages. This befitted of course its tool-like quality as something that could be redeveloped at whim for specific tasks.

2.3 The Psychology of Language

In a more finely tuned rhetorical sense, the British empiricists' reconceptualisation of language allowed its proponents to develop a psychological theory of the misuse of language to explain why so many others had such a flawed approach to the natural world and, of course, to pigeonhole these approaches as psychological failure. Words could have a life of their own and as such were usable despite there being no clear meaning or ideas underlying them. According to Locke, when significations are referred to words, not to things, "one may often meet with very clear and coherent Discourses that amount yet to nothing".[22] Bacon pushed this strongly: "For while men believe their reason governs words, in fact, words turn back and reflect their power upon the understanding and so render philosophy and science sophistical and inactive".[23] He also saw the perpetuation of the syllogism as more or less psychological:

> And no wonder, for it is a thing most agreeable to the mind of man. For the mind of man is strangely eager to be relieved from suspense, and to have something fixed and immovable, upon which in its wanderings and disquisitions it may securely rest.[24]

Words are idols of the marketplace, "the most troublesome of all".[25] While science cannot but rely upon them, their meanings are often already established in common use, and these meanings do not necessarily correspond with divisions in nature. Hence, irrelevant meanings get imported into the study of science, and nature ceases to correspond to theory. Again the culprit was psychology: the mind finding rest in concepts with which it was familiar. Often the mind uses words to make abstractions, as it is want to do, with only a vague idea of what it is the word signifies. Its subsequent employment is thus empty and generative of confusion. Language was thus naturally facilitative of speculative philosophy and hermetic natural philosophy (such as alchemy): "The race of alchemists have built up from a few experiments with a furnace a fantastic philosophy having regard to few things".[26] The particular Aristotelian structure of syllogism only magnified this problem. In this regard, language was vital to understanding the structure and function of science in Bacon's eyes, even if it was also a fundamental source of its problems.

For Locke making the role of language clear in the process of knowledge acquisition was prerequisite to good method. It was thus vital to realize that knowledge is the product of the relation of ideas, not of words. When this fails, however, the first suspect is words themselves. The source of unending disputes in subjects like scholasticism is not the complexity of ideas such fields employ, but the lack of clear significations for their words: "As obscure and imperfect *Ideas* often involve our Reason, so, upon the same Ground, do *dubious Words*, and uncertain Signs, *often*, in Discourses and Arguings, when not warily attended to, *puzzle Men's Reason* and bring them to a Non-plus".[27] Again, as with Bacon, we perceive that speculative philosophy is sustained by words and by the power they have over us. Locke identifies the problem particularly in what he calls "relative Significations",[28] which are referred to other words, not to things. Their relative definitions make them fit to be joined affirmatively and negatively in propositions, not the ideas they refer to: "By this method, one may make Demonstrations and undoubted Propositions in Words, and yet thereby advance not one jot in the Knowledge of the Truth of Things".[29] From this basis Locke can stress caution with the use of hypotheses and "maxims", which, being abstractions, stand at strong risk of possessing no sensible basis or clear ideas. The importance of language to Locke's philosophy thus cannot be downplayed: for him speculation was to a large extent driven by the properties of language, and any philosophical account of natural philosophy had to make explicit the proper role language could have in knowledge production.

Following Bacon, the rhetoric of the Royal Society strongly associated speculative natural philosophy with the psychological dimensions of language, particularly with the ability of the human mind to become absorbed in the "ornaments" of language itself. As Sprat puts it in his *History of the Royal Society*, these ornaments

> are so much degenerated from their original usefulness. They were at first, no doubt, an admirable Instrument in the hands of *Wise Men* when they were only employed to describe *Goodness, Honesty, Obedience;* in larger, fairer, and more moving Images: to represent *Truth* cloth'd with Bodies; and to bring *Knowledge* back again to our very senses, from whence it was at first deriv'd to our understandings. But now they are generally chang'd to worse uses. They make the *Fancy* disgust the best things, if they come sound, and unadorn'd: they are in Open defiance against *Reason*; professing, not to hold much correspondence with

> that; but with its Slaves, *the Passions*: they give the mind a motion too changeable, and bewitching, to consist with *right practice*.[30]

At the end of his history Sprat, like Bacon and Locke, appeals to psychology as the source of risk and for understanding what erroneously drives other disciplines and philosophies.

2.4 Constraining and Instrumentalising Language

In respect of the dangerous psychology of words, one can understand the important emphasis placed by Bacon and Royal Society members on controlling or eliminating the "intervention of words" in order to make language functional for its proper task of communicating distinct ideas. Words were a tool, but they needed to be strictly controlled in their application to the natural philosophical project. Here again the reconceptualisation of language served to aid their particular methodology and exclude other methodologies.

James Stephens in his text *Francis Bacon and the Style of Science* shows how Bacon across his career shaped and formulated a new style of science to avoid the problems of language that plagued the contemporary versions.[31] Whatever the solution would be, it would have to be one that negotiated human psychology and the complex interaction between reason, rhetoric, and the imagination. As we have noted already, Bacon was interested in the creation of a language suitable for science, free of the corruption of the natural languages and of course prior discourses. One option, which shows just how much Bacon distrusted human psychology to reason properly without guidance, would be to rely on languages other than spoken or written language. Bacon was fascinated by the power of hieroglyphs to convey information, as pictures were for Bacon closer to the original sensual impression upon which science began. From pictures indeed a universal scientific language could be constructed: pictures would correspond to meanings and give indications of them so the ambiguity or vagueness of meaning that usually comes with words could be eliminated. According to Stephens, Bacon had two purposes to the reform of method and style: "to shift the emphasis in theory from the author and his manner to the content of discourse; and to provide a method of delivery which is both unpretentious and somewhat veiled".[32]

Bacon was certainly extremely interested in how communication of scientific ideas could take place without distortion of meaning. The scientist should thus strive to present his discoveries and experiences in the same way in which they developed or occurred. In fact a good scientific account would lead the reader to the discoveries the same way the author was led himself. This would not only minimize the ambiguities of language but defeat the reader's natural inclination to take rhetorical style as indication of truthfulness. When we then look at the Royal Society we see similar sentiments.

According to Sprat's history of the Royal Society, the society had resolved to counter the excesses of language in speculative natural philosophy with

> a constant Resolution, to reject all the amplifications, digressions, and swellings of style: to return back to the primitive purity, and shortness, when men deliver'd so many *things*, almost in an equal number of *words*. They have exacted from all their members, a close, naked, natural way of speaking; positive expressions clear senses; a native easiness: bringing all things as near the Mathematical plainness, as they can: and preferring the language of Artisans, Countrymen, and Merchants, before that, of Wits or Scholars.[33]

The Royal Society had picked up the spirit of Bacon's refinement of language for their purposes. A member of the Royal Society who has come in for especial examination in this respect, as a result of the work by Shapin (1984), is Robert Boyle. As Shapin states, "The foundational category of the experimental philosophy, and of what counted as properly grounded knowledge generally, was an artifact of communication and of whatever social forms were deemed necessary to sustain and enhance communication".[34] Shapin's study of what he calls Boyle's literary technology and linguistic boundaries has relevance to us. But we need not understand these necessarily as nothing more than a set of rhetorical devices for carving out authority or constructing an audience, as Shapin seems to suggest. This is not enough to explain why Boyle thought it necessary to construct language the way he did or why he thought his particular methods would be persuasive. Boyle was, like Bacon, perceptive of the plague that afflicted all varieties of speculative natural philosophy:

> Your hermetic philosophers present us, together with divers substantial and noble experiments, theories, which either like peacocks feathers make a great shew, but are neither solid nor useful; or else like apes, if they have some appearance of being rational, are blemished with some absurdity or other, that, when they attentively considered, make the appear ridiculous.[35]

Boyle was working in a context where language was a suspicious and problematic entity which could not be taken for granted without any explication of its proper use. Shapin seems leave this aside, arguing that the purpose of Boyle's literary technology was to multiply effective witnesses to an experiment and thus accord it legitimacy. This is no doubt true, but in this anti-speculative environment, showing the control and subordination of language itself through plain styles and reporting was also likely fundamental to legitimacy. Boyle was constructing his language to eliminate what was perceived to be the improper excesses and ambiguities of the past. He showed a strong desire to recast the reporting of observations and experiments in a non-rhetorical manner that conveyed all the details of the experiment itself, thus eliminating the influence of language in the form of metaphor, false meanings, or ambiguity on scientific communication. If successful, the project would allow the experimenter to proceed in the task of accumulating knowledge without fear that language would in any way hijack the process.

Locke also presented a theoretical dimension to why language needed to be controlled or disciplined if it were to be functional. Again he gave theoretical depth to Bacon's own reservations. In his words, "If we consider, in the Fallacies, Men put upon themselves, as well as others, and the Mistakes in Men's Disputes and Notions, how great a part is owing to Words, and their uncertain or mistaken Significations, we shall have reason to think this is no small obstacle in the way to Knowledge".[36] Words for him were intrinsically "imperfect" because they were separated from the world and limited in their capacities.[37] The world itself is complex, and the best we can do methodologically is operate with clear definitions and simple ideas of the kind we derive from natural historical studies with emphasis on particulars rather than generalities. Resorting to complex terms in the form of substances, purportedly picking out real essences, or mixed modes (abstractions of groups of simple ideas), often in reality refers to nothing:

> Mixed Modes, therefore, that are made up but of a few and obvious simple *Ideas*, have usually no very uncertain Signification. But the names of *mixed Modes*, which comprehend a great number of simple *Ideas*, are commonly of very doubtful and undetermined meaning . . . The names of Substances being annexed to *Ideas* that are neither the real Essences, nor exact Representations of the patterns they are referred to, are liable to yet greater Imperfection and Uncertainty, especially when we come to a philosophical use of them.[38]

Yet notions like "substantial form", "vegetative soul", "abhorrence of a vacuum", and "intentional species" were standard amongst scholastics.[39] It was naturally presumed that such terms referred to real essences that bind the things so referred to. This only served, however, to give a word a putative independence of its confused or uncertain referent:

> In mixed modes it is the name that ties the combination of simple ideas together, and makes it a species. The near relation that there is between species, essences, and their general name, at least in mixed modes, will further appear when we consider, that it is the name that seems to preserve those essences, and give them their lasting duration. For, the connexion between the loose parts of those complex ideas being made by the mind, this union, which has no particular foundation in nature, would cease again, were there not something that did, as it were, hold

> it together, and keep the parts from scattering . . . How much, therefore, in mixed modes, the unity necessary to any essence depends on the mind; and how much the continuation and fixing of that unity depends on the name in common use annexed to it, I leave to be considered by those who look upon essences and species as real established things in nature.[40]

In these respects then the reconceptualisation of language as a tool for reporting and communication required practical constraints on natural history and natural philosophy that suited the agenda of the British empiricist movement. It required the plain style of reporting and the "interpretation-free" documentation of events. To do otherwise was to subvert the function of language and thus give up the ability to give an account of the world.

3. CONCLUSION

These days it is natural amongst practicing scientists to believe in the distinction between sign and thing represented, to the extent that no one much thinks about it. It has become an internalized part of science, as have many of these literary technologies such as the preferability of the "plain style" and "objective" reporting of experimental events. Even when nationalistic questions have come to surface, as documented by many in this volume, it has not necessarily been to question these representational and communicational aspects of language in science, but rather to control them. In some respects, the fact that language has no special ontological status in this regard has perhaps helped facilitate the ability of national or cultural groups to demand science be linguistically nationalized. After all, what can be done in one language can be done in another.

In any case, few in science give language the exalted status it once had, excepting perhaps those physicists or mathematicians who see mathematics in Platonic terms as the language of nature. Perhaps this may be the only kind of modern viewpoint that comes close to the "emblematic world view" of the Renaissance. For the vast majority, language is treated as tool for representation and communication. As we have seen here, this was by no means obvious in the seventeenth century and required concerted development into a fully fledged linguistic philosophy at the hands of the British empiricists and the Royal Society. For these natural philosophers, reconceptualising language as a tool served a number of methodological and rhetorical ends and thus underwrote their empirical and experimental turn in the context of hostile established traditions. This was a dramatic change, one that is ultimately of essential importance for understanding how and why the empirical turn took place in Britain in the seventeenth century.

NOTES

1. In much of part one I owe a considerable debt to Ashworth's "Natural History and the Emblematic World View" for its documentation of the transformations undergoing natural history at the turn of the seventeenth century. William B. Ashworth Jr, "Natural history and the emblematic world view", in *Reappraisal of the scientific revolution* eds. David C. Lindberg and Robert S. Westman (Cambridge, Cambridge University Press 1990) 303–32.
2. Alexander Top (1603), *The Oliue Leaf* (A4v), reprinted in *English Linguistics, 1500–1800: A Collection of Facsimile Reprints,* no. 302 (Menston, Yorkshire: Scolar Press, 1971).
3. Edward Topsell (1607), *A History of Four-Footed Beasts* (A5v), Sidney Warhaft, "The Providential Order in Bacon's New Philosophy", *Studies in Literary Imagination* 4 (1971): 49–64.
4. John Locke, *Essay Concerning Human Understanding*, ed. Peter H. Nidditch (Oxford: Clarendon Press, 1975), 3.II.1; p. 405.
5. William B. Ashworth Jr, "The Revolution in Natural History", in *The Scientific Revolution*, ed. Marcus Hellyer (Maldon: Blackwell, 2003), 132–156.
6. François Jacob, *The Logic of Life: A History of Heredity*, trans. Betty E. Spillman (New York: Vintage Books, 1976).
7. Ulisse Aldrovandi, *Ornithologiae II* (Bologna, 1600), pp. 1–31.
8. William B. Ashworth Jr, "The Revolution in Natural History", in *The Scientific* Revolution, ed. Marcus Hellyer (Maldon: Blackwell, 2003), 132–156.
9. Martin Elsky, "Bacon's Hieroglyphs and the Separation of Words and Things", *Philological Quarterly* 63.4 (1984): 449–60.
10. Ashworth, "The Revolution in Natural History", 148.
11. William Rawley, "To the Reader", in Francis Bacon, *Sylva Sylvarum*, in *The Works of Francis Bacon, 14 vols.*, eds. James Spedding, Robert Leslie Ellis and Douglas Denon Heath (London: Longman, 1857), vol. II, 335–7 (p. 335).
12. Michel Foucault, *The Order of Things: An Archaeology of the Human Sciences* (New York: Random House, 1970).
13. Jacob, 28–29.
14. Locke, II.8; p. 408.
15. Locke, 3.IV.19.
16. Bacon, 1.54; p. 62.
17. Francis Bacon, *Novum Organon*, trans., Peter Urbach and John Gibson (Chicago: Open Court, 1994), 1.14; p. 46.
18. Locke, 4.XVII.4; p. 675.
19. See Anstey, in which the relation between Locke and Baconian method is made out very clearly. Peter Anstey, "Locke, Bacon and Natural History", *Early Science and Medicine* 7.1 (2002): 65–92. See also by the same author, "Locke on Method in Natural Philosophy", *The Philosophy of John Locke* (London: Routledge, 2003).
20. Francis Bacon, *De Augmentis Scientiarium*, trans. James Spedding, Robert Leslie Ellis and Douglas Denon Heath in *The Works of Francis Bacon, 14 vols.*, eds. James Spedding, Robert Leslie Ellis and Douglas Denon Heath (London: Longman, 1857), vol. VIII and IX. VI.1; pp. 111–2.
21. Elsky, 455.
22. Locke, 4.VIII.9; p. 615.
23. Bacon, *Novum Organum*, 1.59; p. 64.
24. Bacon, *De Augmentis*, V.4; p. 94.
25. Bacon, *Novum Organum*, 1.59; p. 64.
26. Bacon, *Novum Organum*, 1.54; p. 61.

27. Locke, 4.XVII.13; p. 683.
28. Locke, 4.VIII.9; p. 615.
29. Locke, 4.VIII.9; p. 615.
30. Thomas Sprat, *History of the Royal Society* (Montana: Kessinger Publishing 2003), p. 112.
31. James Stephens, *Francis Bacon and the Style of Science* (Chicago and London: University of Chicago Press, 1975). See also Elsky.
32. Stephens, 87.
33. Sprat, 113.
34. Steven Shapin, "Pump and Circumstance: Robert Boyle's Literary Technology" (1984), reprinted in *The Scientific Revolution*, ed. Marcus Hellyer (Maldon: Blackwell, 2003), p. 76.
35. Robert Boyle, *The Sceptical Chymist* (Mineola, NY: Dover, 2003), p. 227.
36. Locke, 3.IX.21; pp. 488–489.
37. Locke, 3.IX.1; p. 475.
38. Locke, 3.IX.20; p. 488.
39. Locke, 3.X.14; p. 497.
40. Locke, 3.V.10; p. 434.

4 The Beginnings of Scientific Terminology in Polish

Kłos's *Algorithmus* (1538) and Grzepski's *Geometria* (1566)

Jerzy Biniewicz

The beginnings of scientific discourse in the Polish language can be traced to the sixteenth century.[1] It was then that the first scientific texts were written in Polish,[2] namely *Algorithmus: To jest nauka Liczby* (1538) (Algorithmus: This Is the Science of Numbers) by Tomasz Kłos and *Geometria, To jest Miernicka Nauka* (Geometry, This Is the Science of Land Surveyors) (1566) by Stanisław Grzepski. They were intended not only for a specialist audience but also for readers outside the academic community, who were usually not privy to the arcane details of scientific discourse, which at the time included a great deal of hermetic Latin terminology. The changing structure of society, as well as new economic relations that formed during the Renaissance, gave rise to various questions of a practical nature, which in turn boosted the dynamic development of particular areas of science such as surveying, cartography, and astronomy. Researchers, who were not always associated with universities, generally rejected speculative considerations and saw the need for new solutions within the sphere of material culture.[3] They also appreciated theoretical reflection that allowed them to create a holistic picture of the world and to generalize their conclusions. On the whole, the integration of theory and practice led to new models of communicative behaviour in scientific and didactic discourse.[4]

The objective nature of materially based activity gave rise to the continuing need for the making of abstractions from various objects of culture.[5] In the sixteenth century many researchers believed that mathematics was the right tool for solving the kind of practical problems often faced by traders or manufacturers[6] and that, by means of mathematics, abstract models could be constructed and the structure of material practice made accessible. Arithmetic and geometry contributed to the improvement and acceleration of economic processes. The didactic-scientific discourse allowed the combination of academic abstract thinking with concrete practical solutions and became an instrument for the deep understanding of the structure of events and the world of material culture. For instance, Thales's theorem was used as an instrument for measuring products of culture (e.g., towers, wells) and nature (e.g., trees). In Poland, the demand for reliable studies in the field of arithmetic and geometry was accommodated by numerous treatises in

Latin on these subjects, which appeared in large numbers at the turn of the fifteenth and sixteenth centuries. Amongst such treatises we find *Algorismus minutiarum* (1445), *Nova compilatio algorismi minutiarum* (1447), and *Geometriae practicae seu Artis mensurationum Tractatus* (c. 1450), all three by Marcin Król of Przemyśl. This chapter will consider the next generation of works of this kind: Tomasz Kłos's *Algorithmus* (1538) and S. Grzepski's *Geometria* (1566), landmark studies in the fields, respectively, of mathematics and geometry.

1. *ALGORITHMUS*: THE FIRST HANDBOOK OF ARITHMETIC IN POLISH

The first handbook of mathematics in Polish was *Algorithmus* (1538), by Tomasz Kłos, which contained the basics of merchant arithmetic. Tomasz Kłos gave Polish traders and craftsmen an insight into the rules of arithmetic that substantially helped them carry out economic activities, providing them, for instance, with algorithms for converting various units of measure.[7] At the time, several dozen different units of measure were used in Krakow, and their conversion was a major difficulty in everyday business dealings. Kłos's treatise was not unique in sixteenth-century Europe;[8] indeed, during the fifteenth and sixteenth centuries a number of handbooks on arithmetic were published in different countries, such as *Das Bamberger Rechenbuch* by Ulrich Wagner (1483), *Divina proportione* by Luca Paciali (1509), *The Ground of Artes* by Robert Recorde (1543), *Rechenbuch auff Linien und Ziphren* by Adam Ries (1578), and *Arithmetica oder Rechenbuch auff den Linien und Ziffern* by Caspar Thierfelder (1587). They were addressed to tradesmen and manufacturers and contained arithmetic rules and formulas by means of which various calculations could be made in a relatively short time.

The appearance of mathematical algorithms in sixteenth-century Poland should also be considered from a broader perspective, in the sense that in the world of material endeavours no identical or standardized objects existed.[9] It was thus particularly difficult to calculate the value of different products and compare them, since in each province in Poland there were different systems of weight and measures. This undoubtedly had a significant impact on the authors of the first texts on mathematics written in Polish; they understood how important it was to establish a single common system of weights and measures for the whole community and were convinced that the characteristics which could be described by means of one uniform metric system could be comparable. In the absence of such a generally approved system of measures, any economic and mathematical operation was arduous. In *Algorithmus* by Tomasz Kłos, several dozen units of measurement can be found, all of them used in sixteenth-century Krakow and, more broadly, in Central Europe in the sphere of material practice.

Kłos justified the necessity of introducing arithmetic algorithms into material practice by writing about the elaborate and complicated system of measures by which 1 pound (in weight) was equivalent to 32 litas, 26 pounds to 1 stone, 5 1/2 stones to 1 hundredweight (or centum weight). Whereas the process of converting measures seemed easy,[10] it was in fact rather complex and laborious. For example, in order to establish the value of 6 1/2 zlotys, 14 wierdunek, 19 grosz, and 20 pieniądz, one had to perform the following operations: 20 pieniądz is 1 grosz and 2 pieniądz, 6 1/2 zlotys is equal to 30 grosz multiplied by 6.5, which is 195 grosz, 14 wierdunek is 168 grosz. After adding 20 grosz to this, we get 383 grosz and 2 pieniądz. If 48 grosz is 1 grzywna then 383 grosz is equal to 7 grzywna, 1 zloty, and 2 pieniądz.[11] Similar commentaries associated with converting units of measure can be found in other sixteenth-century treatises, such as S. Grzepski's *Geometria* (1573), which discusses the following system of interdependent measures: "Fifteen ells make one *laska*, three *laskas* make one *sznur*, four *sznurs* make one lan in breadth, and in length there should be 90 *sznurs*".[12]

Some writers noticed the following interdependence: if a measure is isolated and becomes an independent description, that is, an abstract relation between objects (which can afterwards be expressed as a quantity), then products of nature and culture can be described in mathematical terms and captured by algorithmisation (i.e., abstract models created by science). For instance, *regula detri* (equating elements by two equal fractions, e.g., *b*, *c*, and *d*), which is also mentioned by Tomasz Kłos, allowed mathematicians to show the universal a/b = c/d, such that one element is given by the knowledge of the three other elements through the dimension of the rules of proportion. By introducing the rule that any given unit of measure is a pure number (e.g., a measure of volume, weight), which can be put into an equation, mathematicians were able to use the abstract mechanism of quantitatively calculating various values of given measures. Similarly, finding the surface area of a square or a rectangle was performed according to one formula; one side of the figure was multiplied by another (see Grzepski's *Geometria*). Units of measure were a matter of convention, which did not change the fact that a geometric figure, independently of the unit of measure, had to be described by means of the same universal method through a universal geometrical formula.

As early as the fifteenth century, Marcin Król, a professor at Krakow University, sustained in his *Geometriae practicae seu Artis mensurationum Tractatus* (c. 1450) that mathematics was an effective instrument for managing material practice because it presented the ever-changing and non-standardized world of material culture in terms of algorithms or abstract models:

> There are two main branches of geometry, namely theoretical and practical. The theoretical branch deals mainly with quantitative relations by means of reasoning, considerations in our mind only—the thing though

> delightfully great and most perfect but also exceedingly difficult if, according to the philosopher [Aristotle] in the treatise entitle *Posteriora*, all concepts and ideas must preceded by cognition and perception. The aforementioned branch of geometry should be preceded by the practical branch, where by analyzing quantities already known to us, we can have an insight into quantities less known to us. The known quantities can be the following: span, ell, inch, foot, passus, stadium, new and old mile . . . The establishing of these values is called measurement; of these measurements some are artificial, others natural.[13]

In Kłos's *Algorithmus*, common experience of the world is presented with a new vision of reality described in a new manner by means of terminology reflecting arithmetic. Kłos wanted to explain to his readers the concept of accurate arithmetic formulae and, therefore, had to introduce them to abstract thinking, and thus to acquaint them with a picture of the world in which concrete objects (orts, grosz, marks) were merely units of measure that can undergo algorithmisation. Also, he had to familiarize them with arithmetic terminology, as well as the terms that form particular links in the conceptual system of arithmetic and become the exponents of mathematical operations. In order to achieve this, Kłos took into consideration the previous knowledge of his readers and, as a result, decided not to use Latin terminology but rather their native vocabulary, with clear etymologies. In this way he made the abstract terms more familiar and easier to understand. Accordingly, the Latin term *fracta* (currently, Polish *ułamek*, English *fraction*) was replaced by Kłos with "broken number" (Polish *liczba łamana*), in that it consists of two components: the first one called "numerator" (Polish *licznik*, from *liczyć*, see *numerator*), the second one "denominator" (Polish *mianowacz*, form *mianować*, see *denominator*):[14] "A broken number is recorded with twofold figures: the upper one and the bottom one, and there is a stretched line between the figures. The one on top is called numerator and the bottom one denominator".[15]

The readers of *Algorithmus*, who were forced to wrestle on a daily basis with various measures and a myriad of monetary units, discovered through Kłos a new manner of conceptualizing the world. The algorithm proved a universal instrument to control or regulate material practice. Arithmetic as used by Kłos became a source of cognition and a methodologically reliable and epistemologically credible tool for describing the material world.

2. *GEOMETRIA*: THE FIRST HANDBOOK OF GEOMETRY AND LAND SURVEYING IN POLISH

Another book that attempts to combine mathematical abstraction and theoretical models in everyday economic practice is S. Grzepski's *Geometria* (1566), the first Polish work on Euclid's *Elements*. Stanisław Grzepski wanted to acquaint the readers of his *Geometry* with the basics of Euclidean geometry. *Geometria* is not, however, an accurate translation of *Elements*; the first twenty-five pages, for instance, are a pithy description of the first four books by Euclid. Grzepski was aware that the academic community knew *Elements*, for the first three books were widely commented on at Krakow University in Latin. However, Grzepski wished to address practitioners (craftsmen, builders, land surveyors, landowners), who were not participants in the scientific discourse and who needed concrete tools for measuring objects and conducting their economic activity. Grzepski used Latin or Greek terms interchangeably, as well as their Polish equivalents, throughout his work. He was well aware that his text was for practitioners or students who were not necessarily fluent in Latin. Moreover, *Geometry*, being a didactic text for practical builders or land surveyors, used data and facts known to such practitioners of the time. Accordingly, it was impossible for Grzepski to eliminate entirely the vernacular vocabulary used by most of his readership.[16] In fact, he believed that the vernacular should be the key means in the process of forming a didactic-scientific discourse in Poland:[17]

> Therefore, in my aspiration towards stirring our nation into gaining this knowledge, I have written these little volumes in Polish. There were formerly many writing about these matters, particularly Euclid, an ancient Greek scholar, who is even nowadays much treasured by contemporary learned folk. However, I have completed the textbook in Polish so that it could be effortlessly comprehended by everyone. My writing is brief and concise: firstly, so that everyone can briskly grasp and bear the contents in mind; secondly, I heeded the philosopher Socrates' advice to study as much geometry as is needed for these common things such as surveying fields . . . Following this very advice, I have written my books, not for those who do nothing but pore over large volumes.[18]

Grzepski proposed that the model for describing the world by means of language allows us to turn mathematical objects derived from the observation of nature into an effective tool for managing economic practice. Consider his reflections on measuring the area of variously shaped and irregular surfaces:

> The figure *Rombus* can also be effortlessly measured by means of a different method. Do as follows: Measure the *Diametrum* from one angle to another, which will cleave the *Rombus* in twain. Afterwards take the second *Diametrum*, on the opposite side, so that the two *Diametrums* cross each other. Subsequently, find the measures of the two *Diametrums* and then take half of the first one by the whole length of the second one. Here follows an example: If one Diameter has six ells in length and the second one eight, use half of the first one and the whole second one, multiply three by eight or four by six and you shall obtain four and twenty.[19]

Grzepski also proposes that only an abstract and methodologically reliable research project allows us a profound insight into the complex world of nature. He wrote in the preface to his work that "Geometry is the principal point of all sciences":

> Those who have fathomed it thoroughly know how large our Earth or the Moon are, that the Sun is much greater than Earth or Moon, how far is the Moon from the Earth or how far is the Sun from the Moon, how far is it to heaven and how great is its periphery. Those wise people learned about all that through geometry, which is a science truly infallible and solid.[20]

For Grzepski, it was particularly important that Euclidean geometry as a mathematical object proved the divine order of the world, God's omnipotence, and the perfection of the Creator's design.[21] Thus, by studying Euclid's works we can glean an insight into God's design, as the world is created by God in such a way that it can be measured by means of tools presented by Euclid: "Plato when enquired what God does, responded in Greek words . . . well, he plays with geometry".[22]

When discussing the order of the world discovered by science, Tomasz Kłos pointed out that it could be described by means of figures, which by being marks of the number of objects or multipliers of units of measurements are also categories that help us discover the logic and harmony of God's design, that is, the order of the world:

> It is good enough to use our brain to realize that figures are respectable and helpful science. Their respectability can be clearly seen through the fact that all has been created under certain numbers. The Creator of All, too, never gave up using numbers in His grand design, since one is the first of all subsequent numbers; likewise, God, who initiated all objects and creatures, desired to remain within the unity of all existence and have no equals to Himself . . . This is why we have one God, one world, one Sun, one Moon . . . Afterwards comes the Trinity, three being the first uneven number, perfect in itself, which comprises the beginning, the middle and the completion, and all other perfect numbers can be derived from it (six, nine etc.).[23]

While writing his book on the basics of Euclidean geometry, Grzepski was forced to make use of the Latin and Greek terminology that was effective in the academic discourse of the time. In the first part of the work, readers find units of nomination (loanwords, loanwords-quotations) derived from Euclid's *Elements*. These include *punkt* (Latin *punctum*, English *point*), *linea* (Latin *linea*, English *line*), *linea perpendicularis* (Latin *linea perpendicularis*, English *perpendicular line*), *angelus planus* (Latin *angelus planus*), *superficies* (Latin *superficies*, English *surface*), *kathetos* (Greek *kathetos*, English *vertical*, *upright*), *skalenos* (Greek *skalenos*, English *uneven*), *trapezium* (Latin *trapezium*, English *trapezoid*). Also, every time Grzepski used a foreign term, he informed his readers which language he had borrowed it from: "And such line is called *kathetos* in Greek, and *perpendicularis* in Latin".[24] The introduction of foreign terms into the text was intentional because it lent credence to the content and message. Moreover, Polish readers could be instructed on the terminological system used in academic discourse. Foreign terms were explicit and unambiguous, and they did not trigger any undesired associations. Furthermore, they constituted positive proof that Polish treatises were part of the European discourse; foreign lexis gave colour to the message and let Polish readers see Polish texts through the prism of European science.

Conventional and schematic syntax also served a function, as repeated formulas tend to announce new terms and indicate that a definition is going to follow, for example:

a) *X is . . . which . . .*

> A linea is not straight which goes from one point to another and does turn to one side on the way . . . A linea is circular which goes around and after making a circle meets the point where it starts. A linea is bent which goes from inside and in a circle to the top. A linea is curved which goes on turning to one side or another on the way.[25]

b) *When X goes across . . . , then . . .*

> When one straight line goes across another straight line . . . then the two resulting angles are right . . . When a line travelling across another line makes unright angles, then one angle will be pointed and the other one will be thick . . . When a straight line . . . goes to the other side . . ., wherefore the four angles that come into being in that way are either straight or equal to four straight angles.[26]

Another clear means of introducing foreign language terms into the reasoning is the use of a different typeface (italics or capitals): "*Perpendicularis* line, or Kathetos (as I said) is the one which".[27] Scientific terms are also set apart in the text by combining them with drawings or illustrations, which can be considered a parallel narration field, one which helps an unskilful reader understand a complex abstract plan of an academic nature. Grzepski follows Euclid's steps here, connecting geometry with drawings. On many occasions his logical steps are not purely verbal, but presented by means of geometric figures: "One does not need to dwell on proving this, since it is good enough to take a closer look at the drawing".[28] Combining a scientific term from Latin (which would have an unintelligible etymology for average Polish speakers) with drawings or symbols is a functional procedure which helps an unskilled reader access a cognitively complex discourse.

Latin or vernacular compound words, that is, combinations of two or more lexical units, function meaningfully in Grzepski's work as useful instruments of notation. Their key role is confirmed by the fact that they appear in names of geometric figures: "And the angle that lies between two straight lines: one is right and the other one is acute"; "And when a straight line crosses another straight line, and the resulting angle is not right, then one angle is acute and the remaining one obtuse".[29]

Terminological clusters also appear in the second part of *Geometry*, where Grzepski discusses units of measurement used in Poland at the time. For example, consider in the following extracts Polish *chełmieńska włóka* and Franconian *łan, pręt miernicki*: "*Włóka* is a Prussian measure; this is why it is called *chełmieńska*. Similarly, *łan* is called Franconian; it must be then a French unit of measure, it must have come from France"; "Land surveyors in Mazovia use the 'merchant's ell' which is used in that district. A merchant's rod is a quarter of merchant's ell".[30] The vast majority of compounds in the second part of Grzepski's book are two-word clusters made up of a noun (describing a broader concept) and an adjective, another noun, or a participle. This is the case with "square rope" (Polish *sznur kwadratow*, ca. 75 ells), the angle of a "wedge" (Polish *kąt klina*), and an inscribed square (Polish *kwadrat wpisany*). In order to make his text clear and comprehensible, Grzepski used words from the general lexis. As a result, elaborate relations were easier to recognize for readers who had no regular contact with basic academic discourse. A book on geometry with multiword terms containing words from the general vocabulary thus reassured readers in their conviction that popular cognitive experience is not contrary to scientific cognition.

Multiword terms with words from the general lexis appeal to popular associations and offer a vivid account of the complexity of the conceptual scheme of the theory. Hence, as noted earlier, Grzepski distinguishes between the right angle, the pointed angle (nowadays the acute angle, Polish *kąt kończaty*), and the dull angle (nowadays the obtuse angle, Polish *kąt tępy*). The term "pointed angle" (the word *kończaty* is derived from *koniec*, the end, which brings to mind something spiky or peaked) is defined by Grzepski in opposition to the dull/obtuse angle (Polish *kąt tępy*; *tępy* meaning truncate, flat).

The terms constructed by means of semantic derivation retain some association with their original meanings. The process of abstraction leads to the creation of a lexis that records properties of concrete objects and simultaneously reveals their position in the conceptual scheme, hence mingling two different descriptions of the world. Let us take the example of the wedge (Polish *klin*). In the first part of *Geometria*, Grzepski states:

> The commencement of all figures is the one of three sides, in Latin it is labelled *Triangulum, and we can term it "wedge" (Polish klin). Thus Triangulus* or wedges are not all the same.[31]

The Latin term *triangulum* (nowadays *triangle*) was replaced in the book by a wedge (Polish *klin*), which was often used in the sixteenth-century craftsman's lexis with the meaning of "a wooden nail used for instance for splitting oak blocks or for fastening wooden or earthen constructions". Consider the statement by O. Strumieński, a Master of Fisheries who in his work *O sprawie, sypaniu, wymierzaniu i rybieniu stawów* (1573) (Of Digging, Measuring and Restocking Fishponds) affirms the following:

> Having prepared the sluice in accordance with my instruction, as high as you desire to have it, measure the whole length of it by means of a measuring rod. See to it that it is evenly made. If it is lower on one end and higher on the other, then you can prepare some wood, cut out a wedge, and nail it down fast on the front threshold so that water goes everywhere evenly.[32]

Grzepski took a common word meaning a wooden nail used in craftsmanship and, through a metaphor, let an unskilled reader have an insight into the complex world of scientific considerations. As a result, Grzepski initiated the mechanism of abstraction, turning a material object into a mathematical one. Therefore, semantic derivation is in Grzepski's project an important (because functional) means of generating terms for scientific concepts.

Grzepski's handbook, an analysis of real objects described in abstract terms, gives readers an effective tool for measuring existing objects and turns properties of abstract mathematical objects into properties of material objects to be then efficiently measured by geometrical analysis. The natural postulates of Euclidean geometry have not lost their original (popular) sense and have become common ground for reflection, combining the theoretical model with the material reality.

Conclusions drawn from popular experience of the world, realized by means of vocabulary from the general lexis, have become components of the theory without losing their original literal meaning. There is no contradiction between theory and practice; indeed, abstraction is naturally coupled with reality, which is carefully observed: reaching a theoretical reflection is achieved by abstracting formal relations in the material world. In Grzepski's *Geometria* readers will find mutually complementary descriptions in which particular elements in the conceptual system of the theory have their equivalents in the form of lexemes recording the popular vision of extralinguistic reality. For example, Grzepski uses the native words *zwierzchność* (surface, appearance) and *równia* (plane) for Latin *superficies*, in that a plane can be imagined as a sheet of paper or a flat field (the Latin term *superficies* literally means *surface*). Therefore, the flat surface of the earth (Grzepski use two words interchangeably: *plane*, Polish *równia*, and *ground*, Polish *ziemia*) is classified as *superficies*. Also, the words *cane*, *tower*, and *shadow* have their equivalent: *straight line*. Hence, a *cane* meaning a stick resembles a *line*, and the same can be said of *shadow* and a *tower*. Thus, the properties of a cane make it similar to an ideal geometrical object, and a person unskilled in reading geometrical concepts would easily associate the words *cane*, *tower*, *shadow*, and *straight line*:

> Whenever four things have the same ratio, as the first one to the second one, the third one to the fourth one, then accordingly all will have the same ratio . . . For instance, we have four things: a small shadow—first, cane—second, a large shadow—third, a tower—fourth, they all have the same proportions.[33]

Grzepski turns abstract mathematical objects into instruments for describing the real world. The readers of *Geometria* can therefore see the connection between theoretical and practical objects, the abstract and popular perception of the world complementing each other. In other words, defining mathematical objects in the theory does not come into conflict with describing properties of objects in material practice.

3. CONCLUSION

An analysis of sixteenth-century texts in Polish on the basics of bookkeeping and geometry reveals a vision of the world that blends intersecting discourses: theological, scientific, and didactic. The discourses are the result of popular, common-sense, and intuitive approaches to the world,[34] and also scientific (or proto-scientific) ones, in the sense that empirical procedures and *a priori* models become markers for truth and verifiability. The first authors of scientific treatises in Polish understood that Latin and Greek terminology used in European academic or scientific-didactic discourse needed to be replaced by vernacular lexis if they wanted to get their message across. In the sixteenth and seventeenth centuries, Tomasz Kłos, Stanisław Grzepski, Olbrycht Strumieński, and Stanisław Solski argued that the Polish language was as effective for scientific communication as any other tongue, classical or vernacular.[35] Moreover, scientific vocabulary in the vernacular was seen as an expression of the autonomy of Polish academic discourse, a token of prestige and the confirmation of Polish intellectual potential.[36] Polish writers and artists of the Renaissance gradually adopted Polish, claiming that a great nation must create culture in its own language.[37] However, in the sixteenth and seventeenth centuries, science and academic discourse were still bilingual (Polish-Latin).

The system of communication used in academic discourse, which was itself being formed at the time, greatly influenced discursive strategies.[38] The system led to an increasing awareness among sixteenth-century scholars, practitioners, and teachers of the autonomy of scientific jargon within the general language and stimulated the use of jargon and the uniformisation of special lexis. New systemic solutions for creating scientific terms heralded the appearance of mechanisms eliminating the language that could not be used for showing an abstract model of the world. Naturally, folk terms and traditional-professional terms in Polish didactic discourse survived, since scientific and scientific-didactic texts describing the world of abstraction made use of vocabulary from the sphere of material practice when they referred to activities such as crafts or trade. It is thus not surprising that traditional or professional vocabulary appears in seventeenth-century texts, for instance, in the fundamental work for Polish academic-didactic discourse, *Geometra Polski* (1683–1686) by S. Solski.[39]

An analysis of the first scientific and scientific-didactic texts that appeared in the sixteenth and seventeenth centuries enables us to understand the mechanism of forming a picture of the world. While considering the question of the popular and scientific vision of the world,[40] we must bear in mind that the two have much in common.[41] Especially in its initial phase, when the non-scientific paradigm is still being overcome, popular cognition strategies penetrate science.[42] The world can be described by means of procedures used in popular classification, such as ordering by gradation, analogy, or contrast, and science does not entirely give up the perspective of popular interpretation.[43] Popular cognitive experience does not lead to the creation of a picture of the world at variance with the scientific vision, even if they are not identical. The two narratives or visions intermingle, generating a complementary and coherent vision of the world. Scientific classification is combined with a taxonomy which results from the popular observation of the world.

It can be concluded from the process of formation of academic language in Polish that a language as an instrument of thinking, description, and communication is closely associated with the mechanism of encoding the world, its description and communicative relations. Exposing the social and ideological dimension of a language implies looking at it through the prism of practical conditions. An analysis of the formation of scientific language in sixteenth-century Poland shows certain features that characterize modern Polish academic discourse, namely bilingual communication, a preference for terminological clusters and loanwords as a way of broadening the word-stock of a specialized language, and an awareness of the autonomy of scientific terms, which are crucial for academic discourse.[44] Polish Renaissance scholars wanted to share their knowledge of mathematics and geometry with practitioners and students and to do so drew on European academic and didactic discourse. Through their writing they expressed growing concern about their native language, consciously forming and expanding its word-stock. Polish authors of the first mathematical treatises in Polish were able to combine scientific thinking with a creative attitude towards the Polish language. Typically for the Polish Renaissance, their activities became part of the pronounced tendency towards reforming the language and turning it into a useful tool for describing and analyzing the world.

NOTES

1. Mieczysław Bąk, *Powstanie i rozwój polskiej terminologii nauk ścisłych* (Wrocław: Zakład Narodowy Imienia Ossolińskich, Wydawnictwo PAN, 1984), 60–67.
2. Jerzy Biniewicz, *Kształtowanie się polskiego języka nauk matematyczno-przyrodniczych* (Opole: Wydawnictwo Uniwersytetu Opolskiego, 2002), 58–59.
3. Bąk, *Powstanie*, 38–40. Research in material culture has a long and rich tradition in Poland; it comprises studies on products of material culture and their practical skills. See Tadeusz Poklewski-Koziełł, ed., *Studia nad kulturą materialną wieków od XIV do XVI* (Łódź-Wrocław: Zakład Narodowy Imienia Ossolińskich, Wydawnictwo PAN, 1986); Anna Rutkowska-Płachcińska, ed. *Historia kultury materialnej Polski* (Wrocław: Instytut Historii Kultury Materialnej PAN, 1978), vol. II.
4. Jerzy Biniewicz, "Algorytm Tomasza Kłosa czyli o pieniądzu i arytmetyce kupieckiej słów kilka," in *Monety, banknoty i inne środki wymiany*, ed. Piotr Kowalski (Wrocław: Wydawnictwo Uniwersytetu Wrocławskiego, 2010), 39–50; Jerzy Biniewicz, "Podręcznik jako gatunek mowy," in *Gatunki mowy i ich ewolucja vol. III*, ed. Danuta Ostaszewska (Katowice: Wydawnictwo Gnome, 2007), 61–68.
5. Dirk Jan Struik, *Krótki zarys historii matematyki. Do końca XIX wieku* (Warsaw: PWN, 1960), 126.
6. Alistair Cameron Crombie, *Nauka średniowieczna i początki nauki nowożytnej* (Warsaw: PAX, 1960), vol. 2, 154–157; Eva Germaine Rimington Taylor, *The Mathematical Practitioners of Tudor and Stuart England* (Cambridge: Instytute of Navigation, 1970), 60–64; Jim Bennett, "The Mechanical Arts," in *The Cambridge History of Science: Early Modern Science*, eds. Katharine Park and Lorraine Daston, (Cambridge: Cambridge U.P., 2006), vol. III, 673–689.
7. Algorithms are understood as theoretical models which can be used to describe any system of measure units.
8. Marian Aleksander Baraniecki, *Algoritmus to jest nauka liczby: Polską rzeczą wydana przez Księdza Tomasza Kłosa* (Krakow: Drukarnia "Czasu" Fr. Kulczyckiego, 1889), III–V; Roy Newman, *The World of Mathematics* (New York: Simon and Schuster, 1956), vol. 1, 210–211; Dawid Eugene Smith, *History of Mathematics* (New York: Dover Publications, 1958), vol. 2, 9–15, 190–191; Dawid Eugene Smith, *Rara Arithmetica. A Catalogue of the Arithmetics Written before the Year MDCI with a Description of Those in the Library of George Arthur Plimpton of New York* (New York: Chelsea Publishing Company, 1970), 138–140, 213–219; Dirk Jan Struik, *A Concise History of Mathematics* (New York: Dover Publications, 1987), 80–84; Florian Cajori, *A History of Mathematical Notations. Two Volumes Bound As One* (New York: Dover Publications, 1993), 123.
9. Witold Kula, *Miary i ludzie* (Warsaw: PWN, 1970), 192; Witold Kula, *Teoria ekonomiczna ustroju feudalnego. Próba modelu* (Warsaw: PWN, 1983), 47–50. Wiesław Ozon, *Powstanie fizyki matematycznej* (Opole: Zakład Poligraficzny WSP im. Powstańców Śląskich w Opolu, 1993), 60–63.
10. Biniewicz, *Algorytm Tomasza Kłosa*, 43–44.
11. Biniewicz, *Algorytm Tomasza Kłosa*, 43.
12. Stanisław Grzepski, *Geometria to jest miernicka nauka* (Krakow: Łazarz Andrysowic, 1566) (online: http://winntbg.bg.agh.edu.pl/skrypty2/0225/; accessed 21 June, 2013): "To jest: piętnaście łokci czynią jedną laskę, trzy laski czynią jeden sznur, cztery sznury czynią jeden łan na szerzą, a na dłużą ma być dziewięćdziesiąt sznurów". Here and elsewhere, my translation unless otherwise stated.

13. Witold Więsław, *Matematyka i jej historia* (Opole: Wydawnictwo Nowik, 1997), 265: "Istnieją dwa główne działy geometryi, mianowicie teoretyczny i prak-tyczny. Teoretyczny dotyczy rozważania stosunków ilościowych wyłącznie za pomocą dociekania umysłowego, co jest sprawą jedynie rozumową; rzecz lubo wspaniała i najdoskonalsza, toż i wielce trudna, skoro według słów filo-zofa [Arystotelesa] w traktacie *Posteriora:* wszelkie pojęcie powstaje z po-przedzającego poznania. Zbadanie więc tego działu geometryi należy poprzedzić działem praktycznym, gdzie za pomocą znanych ilości rozumowaniem dochodzimy kolejno do znajomości mniej znanej ilości. Znane zaś ilości by-wają takie, jak: piędź, łokieć, cal, stopa, krok, stadium, mila stara i nowa . . . Wyszukanie zaś tej ilości zowie się pomiarem; z tych zaś po-miarów jedne są sztuczne, drugie naturalne".
14. Tomasz Kłos, *Algoritmus to jest nauka liczby* (Krakow: Cracovia ex Officina Ungleriana, 1538) (online: Polska Biblioteka Internetowa: http://www.pbi.edu.pl/book_reader.php?p=30724; accessed 19 June, 2013).
15. Kłos, *Algoritmus*, 37: "Łamana liczba pisze się dwoistemi figurami: wierzchnią i spodnią, a miedzy nią iest li-nia przeciągniona. Wierzchnią zowią licznik, a spodnią mianowacz".
16. Bąk, *Powstanie i rozwój*, 60–67; Biniewicz, *Kształtowanie*, 73–78.
17. Zbigniew Kuderowicz, *Filozofia nowożytnej Europy* (Warsaw: PWN, 1989), 30–32.
18. Grzepski, *Geometria*, 13–14: "Przetoż ja chcąc naród nasz ku tej to nauce pobudzić, napisałem po polsku ty książki niewielkie. Pisałoć o tym ich przedtym dosyć, a zwłaszcza Euklides starożytny autor grecki, w którym i dziś ludzie nauczeni się kochają. Alem ja po polsku, jako najłacniej mogło być, pisał: aby każdy sam przez się wyrozumieć mógł. A napisałem krótko, przodkiem aby każdy rychlej mógł pojąć i łacniej pamiętać: a druga, żem rady w tym użył Sokratesa filozofa, który radzi geometryej się uczyć tyle ile potrzeba jest do rzeczy oto tych potocznych, jako do mierzenia pól . . . Tej rady używając, napisałem oto ty książki, nie dla tych co nic inszego nie czynią, jedno nad księgami siedzą".
19. Grzepski, *Geometria*, 58: "Może też jeszcze inszym obyczajem figura *Rombus* łacno być pomierzana. I to tak: uczyń *Diametrum* od kąta do kąta, który *Rombum* będzie dzielić na dwoje: i drugi zasię *Diametrum* z drugiej strony od kąta do kąta, tak że ty dwa Diametry na krzyż Pojdą. Potym dowiedz się miary obydwu, wiedzże puł liczby jednego na całą liczbę drugiego. Jako na przykład: jeśli jeden *Diameter* będzie na sześci łokiet, a drugi na ośmi: wiedź puł liczby jednego na całą liczbę drugiego, mówiąc troja ośm, albo czwora sześć, uczyni cztery i dwadzieścia".
20. Grzepski, *Geometria*, 13; "Geometria jest główne miasto wszystkich nauk . . . Nie mianuję inszych na ten czas, którzy przez tę naukę doszli i poznali, jako wielka jest ziemia, jako wielki Miesiąc, a czym większe jest Słońce niźli ziemia, albo Miesiąc. Jako daleko od nas do Miesiąca, a od Miesiąca zaś do Słońca. Jako daleko do nieba, a potym jako wielkie jest niebo w okrąg. Tego wszystkiego oni mądrzy ludzie przez Geometr. Dochodzili, nad którą niemasz pewniejszej nieomylniejszej nauki".
21. Similar views are expressed in numerous scientific texts of the time; see for example Jan Brożek, *Gaeodesia distantiarum sine instrumento & Polybii locus obscurior geometricè explicatur* (Krakow: Ofiicina Andreae Petricovij, 1610).
22. Grzepski, *Geometria*, 8: "I Plato gdy go pytano, co Pan Bóg czyni: powiedział graeckim słowem . . . to jest geometryją się bawi".
23. Kłos, *Algoritmus*, 5–7: "Wszakoż tego i sam rozum dokazuje człowieczy, iż liczba jest nauka barzo zacna i pożytecz-na. Zacność jej z tąd się ukazuje, iż wszytki od początku stworzone pod czystym rządem i pod pewną liczbą są postanowiony. Bowiem i ten który wszytko stworzył: nie chciał się z liczby

wyłamować, gdyż jako jedność jest początek wszelkiej liczby dalszej, tak też on, od którego wszytki rzeczy początek mają, chciał się pod Jednością istności zamknąć, a żadnej równej drużyny z sobą nie mieć . . . A dlatego jeden jest pan Bóg. Jeden świat. Jedno słońce. Jeden Księżyc . . . Potym jest trojca, pierwsza liczba nierówna, wszakoż sama w sobie doskonała, bowiem początek, środek i dokonanie w sobie zamyka i wszystki insze liczby doskonałe (jako jest sześcioro i dziewięcioro etc.) z niej się rodzą".

24. Grzepski, *Geometria*, 23: "I takową lineą po grecku zową kathetos, a po łacinie perpendicularis".
25. Grzepski, *Geometria*, 21: "Linea nie prosta jest, która nie idzie tak do od punktu do punktu, aby się nie miała na stronę unieść . . . Okrągła linea jest, która wkoło idąc, przychodzi do tegoż punktu, od którego się poczęła. Zatoczona jest, która ze wnętrz idzie kołem na wierzch. Krzywa jest, która przedsie idzie unosząc się na te i na owę stronę".
26. Grzepski, *Geometria*, 23–24: "Kiedy linea prosta na drugą prostą przyjdzie . . . tedy ty dwa kąty są proste . . . I kiedy linea prosta przychodząc, nierowne kąty czyni, tedy jeden będzie kończaty, a drugi będzie tępy . . . I kiedy linea prosta . . . przechodzi na drugą stronę . . . , tedy cztery kąty, które się tak uczynią, są albo proste, albo czterym prostym równe".
27. Grzepski, *Geometria*, 25: "Linea *perpendicularis*, albo *Kathetos* (jakom powiedział) jest, która".
28. Grzepski, *Geometria*, 37: "Tu nie trzeba przedłużać dowodząc długo, bo i sam wzrok to ukazuje, iż kąty te które są tam i tam położone, dwa a dwa jeden drugiemu są równe".
29. Grzepski, *Geometria*, 23–24: "Zasię kąt który leży między dwiema prostymi liniami, jeden jest prosty, drugi kończaty, . . . I kiedy linea prosta na prostą przychodząc, nierowne kąty czyni, tedy jeden będzie kończaty, a drugi tępy".
30. Grzepski, *Geometria*, 70–71: "Włóka jest pruska miara: przetoż też ją zowią Chełmieńska Włóka: tak jako łan w Polsce francuskim łanem: przeto ze to jest francuska miara, i z francyjej tu musiała przyść . . . Miernicy w Mazowszu używają łokcia kupieckiego, jaki się zachowa w którym powiecie. Pręt Mierznicki jest na półosma łokcia Kupieckiego".
31. Grzepski, *Geometria*, 27–28: "*I tak początek figur jest, tha co jest o trzech stronach, którą łacińskim język iem* zowią *Triangulum*, a my ją możem zwać Klinem. *Triangulus* tedy albo Klin, nie każdy jest jednaki. Abowiem jeden jest co ma wszystkie trzy strony równe, a taki zową po grecku *Hifopleuros*".
32. Krystyna Kwaśniewska-Mżyk, ed., Olbrycht Strumieński, *O sprawie, sypaniu, wymierzaniu i rybieniu stawów* (Opole: Instytut Śląski, 1987), 173: "Zrąbiwszy ten upust, jakom ci o tym pisał, tak wysoko, jakobyś chciał mieć, przeważysz go synwagą od końca do drugiego końca, patrz równali zrąbiony jest. A jeśliby na jednym końcu był niski, a na drugim wyższy, tedy możesz drzewo przystosować, uciosawszy jakoby klin, a mocno przybić na przednich progach, a to dlatego żeby woda wszędy równo szła, po wszystkiej upuście".
33. Grzepski, *Geometria*, 116: "Iż kiedy cztery rzeczy mają jednaka proporcją, jaką pierwsza do wtorej, taką trzecia do czwartej: tedy na przystęp będą mieć jednaka proporcją [. . .] Jako na przykład: tu cztery rzeczy są, cień mały pierwsza, laska wtóra, cień wielki trzecia, a wieża czwarta, a jednaka proporcją mają".
34. See the relation between the traditional or pragmatic taxonomy, which orders parts of nature generally considered essential for human existence, and the scientific typology. This relation has an impact on the process of deriving names; here we deal with different spheres of the world, which do not exclude one another; on the contrary, they complement one another.
35. Bąk, *Powstanie i rozwój*, 60–68; Jerzy Biniewicz, "Kształtowanie się polskiej leksyki naukowej—mechanizm derywowania pierwszych polskich terminów

matematycznych," in *Żywe problemy historii języka*, eds. Marcin Kuźmicki and Marek Osiewicz (Poznań: Wydawnictwo Poznańskiego Towarzystwa Przyjaciół Nauk, 2010), 47–56.

36. In the second half of the sixteenth century an economically and politically stable Polish-Lithuanian Commonwealth of 8,000,000 people and 800,000 square kilometers enjoyed peace and prosperity.
37. See the well-known quotation by M. Rej, a renowned Polish Renaissance writer: "Among other nations let it always be known/ That the Poles are not geese, have a tongue of their own" (translated by Michał Jacek Mikoś). Rej meant that Polish culture was mature enough to express fundamental national matters in the Polish language.
38. Biniewicz, *Podręcznik*, 61–68.
39. Jerzy Biniewicz, "Początki polskiej terminologii matematycznej," in *Dyskurs naukowy—tradycja i zmiana*, ed. Stanisław Gajda (Opole: Wydawnictwo Świętego Krzyża w Opolu, 1999), 203–209.
40. Teresa Hołówka, *Myślenie potoczne. Heterogeniczność zdrowego rozsądku* (Warsaw: Państwowy Instytut Wydawniczy, 1986), 30–32, 148–149; Ronald Langacker, "A View of Linguistic Semantics," in *Topics in Cognitive Lingusitics*, ed. Brygida Rudzka-Ostyn (Amsterdam: John Benjamins Publishing Company, 1988), 84; Jerzy Bartmiński, "Punkt widzenia, perspektywa, językowy obraz świata," in *Językowy obraz świata,* ed. Jerzy Bartmiński (Lublin: Wydawnictwo Uniwersytetu Marii Curie Skłodowskiej, 1999), 109–127.
41. Jolanta Maćkiewicz, "Kategoryzacja a językowy obraz świata," in *Językowy obraz świata,* ed. Jerzy Bartmiński (Lublin: Wydawnictwo Uniwersytetu Marii Curie Skłodowskiej, 1999), 47–55; Jolanta Maćkiewicz, "Potoczne w naukowym—niebezpieczeństwa i korzyści", in *Językowy obraz świata i kultura. Język a kultura,* eds. Anna Dąbrowska, Janusz Anusiewicz (Wrocław: Wydawnictwo Uniwersytetu Wrocławskiego, 2000), vol. 13, 107–113; Jerzy Bartmiński, *Językowe podstawy obrazu świata* (Lublin: Wydawnictwo Uniwersytetu Marii Curi Skłodowskiej, 2009), 36–37.
42. Hołówka, *Myślenie potoczne*, 61; George Lakoff, *Women, Fire and Dangerous. What Categories Reveal about the Mind* (Chicago: University of Chicago, 1987), 118.
43. Bartmiński, *Językowe podstawy*, 14; Maćkiewicz, "Potoczne w naukowym—niebezpieczeństwa i korzyści", 111.
44. Jerzy Biniewicz, Anna Starzec, "Styl naukowy," in *Przewodnik po stylistyce polskiej*, ed. Stanisław Gajda (Opole: Wydawnictwo Św. Krzyża, 1995), 401–406.

5 Language and History in the Context of the *Société des Observateurs de l'Homme* (1799–1804)

Martin Herrnstadt and Laurens Schlicht

1. INTRODUCTION

In 1799 Pierre-Louis Roederer published an account of the role philosophy played in the revolution. In this account he talked about his contemporaries' place in time: "Philosophy", so he wrote, "is no longer enclosed in the books of the sages [Montesquieu, Voltaire, Rousseau, Diderot and d'Alembert]; she has escaped from them, like the light emanating from the sun; like the light, she is today widely spread throughout the whole world, she shines brightly over all heads; she is reflected by most of the social institutions, mixed with the air which we breathe".[1] Unlike Kant, who, as Foucault argued, understood his age as part of the emerging *Bildung* (education) and *Aufklärung* (enlightenment), Roederer seems to understand philosophy as nearly having fulfilled its mission.[2] Joseph-Marie de Gérando, one of the first to introduce Kant's philosophy in France, argued in a report delivered in 1808 to the emperor that the foundation of reasonable philosophy had been achieved in his own age: "Even if there are still many problems to be solved, the fundamental problems are at least defined with more precision than ever".[3] In an earlier voluminous work, *Des signes* (Of Signs, 1800), he discussed the possibility of reforming language and designing a "perfect philosophical language".[4] In order to do this, Gérando, in various writings, engaged in an analysis of the "filiation" of ideas and signs. He was convinced that "for the filiation of ideas the storehouse of languages will be what the registers of civil status are for the generations of men".[5] The "storehouse of languages", understood in the framework of natural history as the system of classification of languages, provided a powerful basis for the history of the human mind. For Gérando this involved a certain understanding of history: "The secret of the future lies in the past. The history of nations is the first study of the legislator. The history of thought has to be the first study of the philosopher".[6] Gérando was a famous member of the *Société des observateurs de l'homme* (Society of the Observers of Man, SOH). The SOH tried to establish a new empirical paradigm of the science of man that linked the study of human beings closely to the study of language.[7] However, unlike Gérando, Roch Ambroise Sicard, another member of the SOH and the director of the famous school for deaf-mute persons in Paris, thought that a perfect language would be possible here and now. For him, the development of the human mind from a historical perspective was just a collection of failures to be overcome.

The present chapter will thus focus on the question of how language and models of conceptualizing the history of the human mind were conceived by Gérando and Sicard. They both shared what Foucault understood as the classical model of language, that is, the representation of simultaneous sense perceptions in a successive manner.[8] For the members of the SOH language was the medium, as well as the fundamental tool, for organizing society. The intersection of language and society was in fact the starting point of their arguments. They were concerned with topics as varied as deaf-mute education, medicine, the pathology of mental illnesses, and the organization of social institutions known as the *art social*.[9] As Martin Staum has shown, the quest for a reliable and/or a universal language was related to the basic aim of politics in France after Robespierre and the *terreur*. Government institutions like the *Institut national* (of which Gérando and Sicard were members) were to stabilize the French Revolution and to neutralize linguistic controversies (in the past controversial terms such as *nation* or *freedom* had led to fatal misunderstandings).[10] Accordingly, philosophers such as Sicard and Gérando pondered over how a reasonable system of signs could be constructed and whether the dream of social stability could be achieved via a universal system of signs. Yet their proposals differed: whereas Sicard focused on a utopian model and quest for the perfect and natural language, Gérando concentrated upon the administration of the diversity of languages. Still, both developed models of a scientific and empirical nature, and both relied on a specific understanding of the history of the human mind as the necessary precondition for validating knowledge about language. The first two parts of this chapter discuss Gérando's approach to the administration of different sign systems and Sicard's vision of a perfect language respectively. The third connects the different epistemic approaches of the SOH to their respective political agendas.

2. GÉRANDO AND THE DIVERSITY OF LANGUAGES

The various political and epistemological functions of research on language within the SOH relied on several traditions, but mainly on Condillac. At the end of the seventeenth century Locke had formulated basic epistemological assumptions which shaped the framework of the debates on political language and universal sign systems during the French Revolution. The conviction was shared that there were no innate ideas and—more importantly—that, in order to examine the human mind, research on the *origin* of ideas was mandatory. Condillac radicalized these assumptions in the mid-eighteenth century: ideas were not innate, and they were not mental *faculties* either.[11]

For Condillac all mental activities could be explained through the sensations they produced in the mind. Moreover, he argued that every logical operation could be explained by one faculty alone: the faculty of attention. The subject could only perceive its own sensations. Objects were thus only represented through the sensation of them, which meant that objects themselves could not be perceived.[12] If immediate knowledge of the outside world was impossible, then the relation between mind and the outside world became the basic problem of epistemology. If language was dependent on the mind then it was surely arbitrary; indeed, the arbitrariness of signs was one of the fundamental questions debated throughout the eighteenth century. The French Revolution (and above all, the desire to restore order after the *terreur*) renewed the interest in these questions. Roederer, member of the group of conservative republicans responsible for setting a scientific agenda of stability after the demise of the Jacobins, recounts in his papers a long discussion with the "young general" Bonaparte in the spring of 1798 on the question of the relationship of signs and the development of the human faculties.[13]

A fundamental assumption was that the connection between things, ideas, and signs comprised two possible sources of error or abuse: (1) the representation of a thing by an idea and (2) the representation of an idea by a sign. The SOH advocated a scientific approach to these questions, an "experimental" metaphysics based on sense perception.[14] To stabilize the connection between things, signs, and ideas required for the SOH the invention of different systems of observation. To do so the SOH transformed the abstract principles of Condillac into an empirical science based on the collection of as many facts as possible. For example, the same scientific techniques should be applied to the science of history of philosophy (understood as an analysis of the human mind in its temporal dimension), as well as to the observation of savage peoples (the human mind in its spatial dimension).

In his *Histoire comparée des systèmes de philosophie relativement aux principes des connaissances humaines* (Comparative History of the Systems of Philosophy Considered in Relation to the Principles of Human Knowledge, 1804), Gérando sketched out a universal history of philosophy.[15] In the opinion of Wilhelm Gottlieb Tennemann, professor of philosophy at Marburg University and translator into German of *Histoire comparée*, Gérando's was the best history of philosophy in French ever written and Gérando the best connoisseur of German philosophy since Charles de Villers.[16]

In each of his philosophical works, history, as a scientific method of description, was meant to validate knowledge about human beings. *Histoire* was for Gérando an epistemological concept, a basic way to produce knowledge that consisted in the collection and comparison of singular facts.[17] Like natural history in the eighteenth century, the history of systems of thought was the means to turn the study of the human mind into a project of observation and collection of facts. These facts would contribute to the analysis of the changes of the human mind within its development; in other words, history of philosophy was an analysis of the intellectual faculties of man in space and time, and Gérando saw his universal history of philosophy as a method to gather facts for the investigation of the principles of human knowledge. Each philosophical doctrine had its specific "development", as well as a specific "character" and a "physiognomy".[18] These doctrines could not be judged *a priori*, but had to be observed through the facts history of philosophy offered. Moreover, a set of fundamental questions within each doctrine needed to be identified and clearly defined.[19] As a result, each doctrine of this vast *tableau* of the history of philosophy presents a finite number of elements which allow the historian to place it on the historical map—Gérando in fact understood the *tableau historique* in visual terms as a map.[20] Ultimately, the end was to discern a "natural method" (*méthode naturelle*) for the history of philosophy. Since the concept of *histoire* and the methods of writing it constituted a major theme in enlightened botany and zoology, the transfer of the concept of nature to the history of philosophy was not so unusual.[21] What was possibly new in the kind of history Gérando wrote was that in his approach to the history of philosophy and the human mind he combined the French analytical tradition with the concept of *Bildung*, which in philological scholarship described a process of unfolding something that in its essence had always been there (a core). On the one hand, Gérando aimed at the isolation of "elements" in the history of philosophy; on the other, he spoke of "doctrines" that were stable entities. The development of a method that should establish an impartial ground to judge and combine the different systems or doctrines was a major aim of his studies; this he tried to achieve through a specific understanding of *history*, called inductive or comparative. The history of the principles of human knowledge should not only be understood as

> a treatise of philosophy about the first questions of this science, but as a treatise conceived according to the most prudent, but also most neglected, method: the method of experiences. We thus venture to offer it also as an introduction to the history of philosophy or as an essay on *experimental philosophy*.[22]

History as a method of observation or *expériences* (meaning either experiment or experience) differed from discussions around the "history of philosophy" in German debates,[23] and yet Gérando knew and used the abundant German literature on how to write history of philosophy.[24] *Histoire* in Gérando's sense was the claim that every single item of human knowledge could be traced back through the different moments of its generation and simplified to a specific number of basic components, which were in their turn the central concepts of the different branches of knowledge. Alluding to the natural history of Linné, Gérando's historical method aimed to establish "neutral" knowledge[25] and eventually produce a "code of practice *for the administration and application of science*" that would also set out the "rule for a *better regime* of our mind".[26] Gérando sought to compare the different branches of knowledge through a general method and practice of observation. He sought a *code*: a system of general principles that could not exist as theory but had to be applied to each specific case.

In *Des signes*, Gérando elaborated for the first time his concept of administration or harmonization between different sign systems. *Des signes* divided scientific language into four incommensurable sign systems, each constructed upon its specific analytical centre. For Gérando the impossibility of a universal language resulted from human nature itself, which was in his eyes incompatible with the conflicting needs of people inhabiting different circumstances. The aim was rather a general method which allowed the administration of the fundamental differences. Thus revolution turned into a permanent linguistic reform. As he explains in the conclusion of his treatise: "All these reforms brought together will not make disappear entirely the diverse faults which we sense in our languages; they will not meet the requirements of the perfect philosophical language". Also, "by the nature of things themselves", many of the conditions for a perfect language were virtually "incompatible with one another".[27] Thus, the end of any history of the mind was the harmonization of different or even contradictory demands of the sign systems that made up the different branches of human knowledge.[28] A method based on a balance between experience and rational principles would build a ground for a general communication between the nations.[29] To illustrate these ideas, it is worth considering his ineffective travel instructions for the expedition of Baudin to Australia (1801–1803) and his actual practice as an imperial administrator in occupied Catalonia (1812).

In his *Considérations sur les diverses méthodes à suivre dans l'observation des peuples sauvages* (1800), Gérando stated that the philosophical voyager will travel through the ages when travelling to the *peuples sauvages*.[30] In other words, geographical travel meant, under certain conditions, travelling back in time. This type of voyage offered for Gérando a means for the analysis of the human mind and the study of the history of philosophy. The observation of the language of the so-called savages constituted, in this respect, a crucial part of this project.[31] Although Gérando demonstrated a certain interest in the languages of the savages, the core of his project was nevertheless the creation of a development model of civilizations. Through the assessment of the savage's language, Gérando argued the observer could determine the savage's place in the history of the development of the human mind. Gérando, like many of his contemporaries, believed that the savages represented the past epochs of his own stage of development; to them the voyager came as "deputy of the whole mankind".[32] For Gérando, an analysis of language in its components (one of the ultimate goals of the observation of savage peoples) had to begin with the existing state of language development and the description of languages as systems. These systems, which formed coherent entities, were comparable to the doctrines of his history of philosophy as well as to the sign systems described in *Des signes*. Thus the natural history of the human mind was in effect a cultural history veiled by habits, traditions, and customs which had to be analyzed through the prisms of different systems.

Research on language had very practical implications for the practice of Gérando's work as an imperial administrator in the Napoleonic Empire. Gérando was sent to Catalonia at the beginning of 1812, and on the basis of this experience he advocated in a report to the minister of justice that the Catalan language should be part of the legal administration used to implement the *Code Napoléon* in the occupied territories. Unlike the project to suppress local languages in France during the first years of the revolution, Gérando's policy did not aim at linguistic unification but at the integration of local languages into the centralized administrative system of the French Empire.[33] Metaphysical speculations gave way to a certain historical pragmatism in which the political circumstances of Catalonia had to be taken into account. Indeed, Gérando, in his suggestion to the minister, clearly saw that linguistic and cultural diversity could be turned into a powerful tool of government:

> Not for the reason of being understood have I submitted to *your Excellency* the idea to grant the Catalan language the right to be the language of the public acts. I presented this idea only in a political sense. I believe I mentioned that the use of the Castilian or Spanish language had only been introduced in Catalonia in the time of the submission of this province under the general laws of Spain, of which it had been independent until then . . . The Catalan hates the Spaniard, he does not want to be Spanish . . . We can make use of this memory of the past, of these existing antipathies, to make it easier to rule opinion. The choice of the language for the public acts is perhaps one of the most influential things in relation to national opinion.[34]

Gérando's recommendation to the minister of justice is telling. Language was not only a tool to govern peoples, to guide their hatred and affection, but also a tool which could turn history and memory into a resource for the mobilization of public opinion in favour of the imperial Napoleonic State. The Catalonian episode is one of many cases in which Gérando's concern with history was applied to political practice. The dreams of a universal sign system transformed into an instrumental understanding which acknowledged languages within the context of their plurality and as powerful tools and resources of government. In this sense, Gérando's historical method should be understood as a political rationality that underpinned an administrative agenda fundamentally connected with the establishment of a new imperial approach to cultural and linguistic plurality.

3. SICARD'S EXPERIMENTAL PHILOSOPHY OF LANGUAGE

To write the history of the mind meant something quite different for Sicard. Contrary to Gérando, Sicard, "one of [Gérando's] best friends",[35] believed that nature was a kind of language, in the sense that it was a system in which signs represented things.[36] These two systems (i.e., the system of things and the system of signs) had the same grammar. What united the authors of the SOH was that in order to find the perfect language, one had to engage in a kind of research based on gathering facts, thus adopting an empirical approach to the sciences of language. As explained previously this kind of philosophy was seen as an experimental science and thus, it could be argued, as an experimental philosophy of language. Contrary to Gérando, Sicard's approach to this experimental science was devoted to finding the perfect system of signs. As the director of the Paris school for deaf and mute persons, he presented conclusions deriving from his experiences with deaf-mute pupils and hence backed by empirical evidence.

The natural history of the human mind was for Sicard the starting point for the construction of his so-called "methodical sign language", which he claimed to have developed together with his pupils.[37] Sign language (i.e., the language of the deaf-mute) was thought to make visible the true and natural order of ideas, and this truth was in its turn thought to be connected to normativity. As a consequence, the true system would also be a good system. It was only two years after the beginning of the revolution that Talleyrand expressed this conviction in his *Rapport sur l'instruction publique*. Like many of his contemporaries, he connected nature and freedom to the abolition of traditional forms of communication:

> [It is] necessary that the ancient obsequious forms, these timid precautions of feebleness, the suppleness of an oblique language which seemed to fear that the truth would appear entirely . . . will be changed into a simple language, proud and fast; because where thinking is free language has to become prompt and frank.[38]

Deaf-mute language, Talleyrand assumed, was exactly this kind of language, for he believed it was "the living image of thought, . . . by means of which the most abstract ideas become nearly visible, and which, in its decomposition, . . . presents the veritable grammar not of words, but of ideas".[39] In a review Roederer also emphasized the merits of Sicard's *Cours d'instruction d'un sourd-muet de naissance* (1799): "This work will be of supreme utility not only for the deaf-mutes, but for the whole of humanity . . . One will probably find in it a new principle of a perfect pasigraphy and probably even of a philosophical and universal language".[40]

For Sicard (like for Condillac in the *langue des calculs*), a language was a collection of signs designed to enable the formation of judgements which is first constructed by means of a process of giving names to things. In Sicard's general grammar, the development of imperfect natural languages was not controlled by reason, but usually languages were created by sheer coincidence and the accumulation of previous linguistic elements and coinages.[41] Correspondingly Gérando would also admit that for these "unreasonable" languages the method of analysis was difficult to apply. Yet Sicard saw in the deaf-mute a being without a language capable of learning a new language from scratch. The uninstructed deaf-mute was conceived by Sicard as a perfect *tabula rasa* in the Lockean sense of the word, who would fulfil the necessary precondition of his experimental philosophy of language. In the *Cours d'instruction* Sicard affirms that the deaf-mute, unable to communicate, is an "automaton" in need of "some link of communication between him and the other men".[42] This education Sicard describes as

> an education of a savage put into action. During this education, the teacher, because he profits from the very small number of familiar elements of the grammar of this man of nature, will create with him . . . the grammar of the man of society.[43]

If a perfect language was possible, as Sicard affirmed, and if all existing languages were chance products, the controlled education of the deaf-mutes was an ideal means to construct the ultimate language dreamed of by Talleyrand. This language could change the natural taxonomy of things into a taxonomy of signs. In order to form a taxonomy of words (i.e., a dictionary) one first had to know the "subtlest operations of the mind" with a "patience of observation following the example of the anatomist, who considers the most delicate fibre".[44] Sicard thus believed it would be possible to translate a natural taxonomy into an artificial one that fully represented the natural order. For instance, he explained graphically how a sensation (i.e., a simultaneous impression) could be represented in a successive manner in language; by means of the illustration in Figure 5.1, he shows how a red paper (instantly apprehended by the senses) is represented in time, successively. This also implied that, with the aid of language, one could form successive propositions like "the paper is red".[45]

Language was at the same time a means to analyze the civilized state of a human being and a tool to civilize people. According to Sicard, the new elements of general grammar could only be discovered by experiments or experiences (*expériences*) with deaf-mutes.[46] In *Cours d'instruction d'un sourd-muet de naissance* (Course of Education for One Who Was Born Deaf-Mute, 1799), Sicard provides an account of the lessons he delivered to his pupil Jean Massieu.[47] Contemporary German critics claimed that it was not really Sicard who taught the pupils, but his assistant-teachers, and that the *Cours d'instruction* was only a form of philosophical romance.[48] The enthusiastic acceptance of his method in France explains the success of Sicard's public lessons, for which reservations for tickets (as for a performance in a theatre) often were needed.[49] The general excitement over these public lessons arose both out of the perception of the deaf-mute as a hopelessly deprived human being, but also out of the demonstration of the power of education and philanthropy. Sicard even claimed in these lectures that through this collaboration with deaf-mutes he had created new citizens.[50] Sicard was indeed convinced that progress towards a general grammar

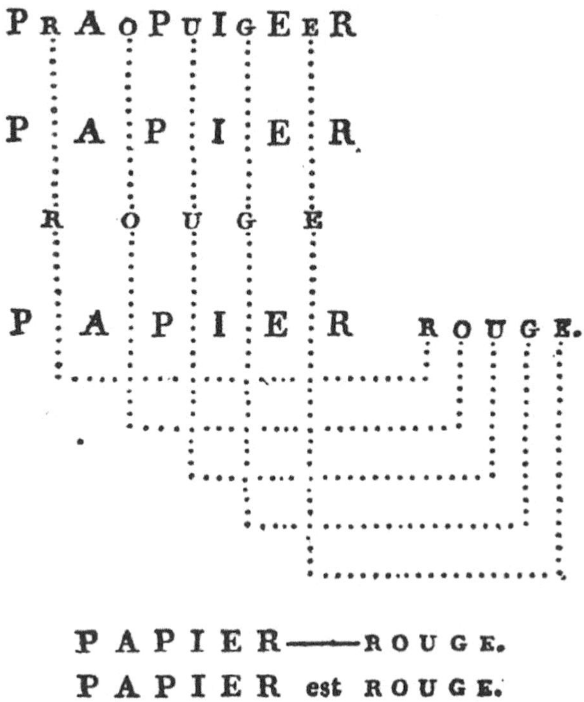

Figure 5.1 Illustration of the function of the judgement (*proposition*) from Sicard's *Cours d'instruction* (p. 46)

necessarily involved an empirical science of language and not merely philosophical reflection. Just like Gérando, Sicard understood language against the background of natural history: "All language, says our philosopher [i.e., l'Abbé de l'Épée, Sicard's teacher and predecessor as director of the deaf-mutes' school], is nothing but a collection of signs. Like a progression of figures of natural history, language is a collection of images, a representation of a large number of objects".[51]

The progress of the deaf-mute is the progress of civilization in a nutshell, with the difference that the teacher of the deaf-mute ensures that this progress is in accordance with the structure of nature.[52] Prior to education, the deaf-mute is a mere automaton, for he is believed to lack developed intellectual faculties.[53] The language of nature, the deaf-mutes' language, could only be regarded as *natural* when a highly *civilized* philosopher, Sicard, constructed it according to civilized concepts of how language should be constructed and with the aid of the tool of a "perfect" grammar.

Sicard made a strong case against the belief that only by learning a spoken language—as opposed to sign language—could one obtain faith, an old idea which was connected *inter alia* to the *fides ex auditu* in the Bible. Attempts to teach deaf-mutes a vocal language not only took a long time, but also had virtually no lasting results. Johann Conrad Amman, for example, summarized in the following way his ideas concerning language (in the English edition used here translated as *voice*): "The *Voice* is an Emanation from that very Spirit, which God breathed in Man's Nostrils, when he Created him a living Soul".[54] The universal system of signs of the reformed education of deaf-mutes was also situated within a specific understanding of politics and time. The history of language and the history of the human mind, Sicard believed, had parallel developments. A natural sign system had to be the basis for the organization of society, and an empirical science of language was the means to achieve that.[55] To conceive of the deaf-mute as a *tabula rasa* translated universalistic conceptions on language into empirical observation. By adapting to new empirical forms of knowledge production, Sicard tried to demonstrate the basic insights of Condillac with the aid of facts.

As the head of the so-called *conseil d'administration* (council of administration) and the *conseil de perfectionnement* (council of perfecting [the art of teaching the deaf-mutes]) of the Paris school for deaf-mutes,[56] Gérando organized a collective research program that put an end to the utopian ideal of a perfect sign language and instead tried (by means of statistics, medical investigations, and voyages in which he participated) to establish a factual basis for the ideal arrangement of the deaf-mute's education. In the eyes of Sylvestre de Sacy, part of the *conseil de perfectionnement*, the administrator was the producer of a higher order of knowledge. In a review of Gérando's voluminous *De l'éducation des sourds-muets de naissance* (On the Education of Deaf-Mutes from Birth, 1827), de Sacy compared the work of the administrator in developing the art of the deaf-mute's education with the contribution of the philosopher and the historian: "The former [the philosopher] will engage in fathoming the theory, the latter [the historian] in gathering facts, and the third [the administrator] in seeking in the lessons of theory and the results of experience the means to avoid the pitfalls, to fill in the gaps, to ensure success and to simplify and regulate the methods".[57]

4. CONCLUSION

One of the purposes of the universalization of the laws of language was unification: local practices should be substituted by universal ones. Rationality shared by all would guarantee equal possibilities of political participation. Talleyrand, in his proposal for the reform of the standards of weights and measures, declared that with the abolition of the local privileges of sovereign princes democratization would follow and the peasants would finally profit from this abolition and be emancipated from subordination.[58] Similarly, the reform of language was meant to enable illiterate groups in society to join the path of enlightenment and learning. These approaches attempted to eliminate the *patois* (local languages) and to install general rules for the use of the French language.[59]

These approaches to language were intertwined with a new interpretation of history, one which the *citoyen* authorized and for which he was responsible. The political implications of Sicard's perfect language in the framework of an empirical research design (the deaf-mute's language was in his eyes "a kind of universal language which . . . can be understood by all men of all countries and nations")[60] involved encouragement of general communication and participation in the project. At the turn of the century, Sicard's science was the heir of Condillac's philosophy of language. Unlike Sicard, Gérando gave precedence to the differences between systems of signs and ideas, and so, from his viewpoint, a universal language was no longer an option. His vision of the administrator-philosopher was one who had to harmonize and negotiate between the contradictions existing in reality. Beyond the ideal of the deaf-mute education, his historical method was based on the acknowledgement that linguistic diversity could be a much more powerful tool to govern and unify than direct coercion via a universal language. His methodology and his practical and pragmatic "philosophy of experience" formed the basis of Gérando's career as a state servant throughout the various post-revolutionary political regimes. Gérando was careful to assert that the true philosophy would only strengthen the spirit of obedience because true philosophy teaches respect for the existing order and religion and fights scepticism and atheism.[61]

Sicard's universal tool of communication would likewise work as a means to put an end to controversies. If the deaf-mute was an ideal citizen created by an empirical science of language, there would be no need for change. Between the republican utopia of the birth of a new man with a new language, and the imperial pragmatism of governing a plurality of languages, the SOH is a prime example of how the analytical tradition of the enlightenment could be accommodated to political and social exigencies. In this sense the figure of Gérando is most telling. In his work the science of man transforms from an agent of social change, leading towards equality, to a philosophy of administration. The latter would aim to govern essential differences and constitute a society based on inequality. For Gérando history was the description of the plurality of existing systems of signs, the doctrines of the history of philosophy, and later also classes in society. From the outset of his career as an author in 1800, eradicating differences had not been a desirable goal as it led to violent revolutions and chaos. The enlightened analytical tradition thus offered a model of a comprehensive system of description by which differences could be explained and their systematic place understood. This holds true for the history of philosophy as well as for the administration of the deaf-mute's school. If the secret of the future lay in the quest for the past, as Gérando argued, the wisdom (*sagesse*) of individual and collective destiny went hand in hand with knowing and understanding the advantages of one's place within the existing social order.[62] Gérando translated the eighteenth-century quest for the origin of language and civilization that Sicard implemented in empirical research strategies on the natural history of the mind into a question about the place of language and its development within a far wider and more complex system of circumstances. Given these circumstances language was but one tool (among others) to civilize humankind.

NOTES

1. Pierre-Louis Roederer, *De la philosophie moderne et de la part qu'elle a eue a la Révolution Française* (Paris: Imprimerie de Journal de Paris, 1799), 41. Unless stated otherwise, all translations in this text are our own.
2. Kant defined *Aufklärung* as "man's emergence from his self-incurred immaturity"; Immanuel Kant, "An Answer to the Question: 'What Is Enlightenment?'" in Immanuel Kant, *Political Writings* (Cambridge: Cambridge University Press, 1991), 54–61 (p. 54). Foucault writes in this regard: "Kant defines Aufklärung in an almost entirely negative way, as an Ausgang, an 'exit,' a 'way out.' In his other texts on history, Kant occasionally raises questions of origin or defines the internal teleology of a historical process. In the text on Aufklärung, he deals with the question of contemporary reality alone. He is not seeking to understand the present on the basis of a totality or of a future achievement. He is looking for a difference: What difference does today introduce with respect to yesterday?" Michel Foucault, "What is Enlightenment?" ('Qu'est-ce que les Lumières?'), in *The Foucault Reader*, ed. Paul Rabinow (New York: Pantheon Books, 1984), 32–50 (p. 34).
3. Joseph-Marie de Gérando, "Philosophie", in *Rapport Historique sur les Progrès de l'histoire et de la littérature anciennce depuis 1789 et sur leur état actuel. Présenté à Sa Majesté l'Empereur et Roi, en son Conseil d'état, le 20 Février 1808, par la Classe d'histoire et de la littérature ancienne de l'Institut*, ed. Bon-Joseph Dacier (Paris: Imprimerie impériale, 1808), 278–355 (p. 353): "s'il reste encore beacoup de problèmes à résoudre, les problèmes fondamentaux sont au moins définis avec plus de précision que jamais".
4. Joseph-Marie de Gérando, *Des signes et de l'art de penser considérés dans leur rapports mutuels*, Vol. IV (Paris: Chez Goujon, 1799), pp. 5–6.
5. Joseph-Marie de Gérando, *De la génération des connoissances humaines* (Berlin: George Decker, 1802), p. 29.
6. *Ibid*, 11.
7. Jean-Luc Chappey, *La société des observateurs de l'homme: (1799–1804); des anthropologues au temps de Bonaparte* (Paris: Société des études Robespierristes, 2002).
8. Michel Foucault, *Les mots et les choses* (Paris: Gallimard, 1966), p. 96.
9. Lia Formigari, *Signs, Science and Politics. Philosophies of Language in Europe 1700–1830* (Amsterdam and Philadelphia: John Benjamins, 1993); Sophia Rosenfeld, *A Revolution in Language. The Problem of Signs in Late Eighteenth-Century France* (Stanford: Stanford University Press, 2001).
10. Martin Staum, *Minerva's Message. Stabilizing the French Revolution* (Montreal & Kingston/London/Buffalo: McGill-Queen's University Press, 1996), p. 4, pp. 95–118; Rosenfeld, *A Revolution in Language*, pp. 183–193.
11. See Isabel F. Knight, *The Geometric Spirit. The Abbé de Condillac and the French Enlightenment* (New Haven/London: Yale University Press, 1968).
12. This is a topic widely discussed by Michel Foucault. See, for instance, Michel Foucault, *The Order of Things. An Archaeology of the Human Sciences* (London/New York: Routledge, 2005), pp. 70–74.

13. Antoine Marie Roederer, ed., *Oeuvres du Comte P. L. Roederer [. . .] publié par son fils, Tome troisième* (Paris: Firmin Didot, 1854), p. 323.
14. "*The course of the Instruction of the deaf-mute* will be a kind of *experimental* Metaphysics and Grammar, which can be used for the instruction of all children". Roch Ambroise Sicard, *Cours d'Instruction d'un sourd-muet de naissance pour servir à l'éducation des sourds-muets et qui peut être utile à celle de ceux qui entendent et qui parlent* (Paris: Le Clere an VIII [1799]), x.
15. The most comprehensive overview and bibliography concerning Gérando is in the recent anthology edited by Jean-Luc Chappey, Carole Christen and Igor Moullier, *Joseph-Marie de Gérando (1772–1842). Connaître et réformer la société* (Rennes: PUF Rennes, 2014).
16. Wilhelm Gottlieb Tennemann, "Vorrede des Uebersetzers" (Foreword of the translator), in Joseph-Marie de Gérando, *Vergleichende Geschichte der Systeme der Philosophie* (Marburg: Neue academische Buchhandlung, 1806), pp. vii–xxii (p. x); see also Review of "Histoire comparée des Systèmes de Philosophie considérés relativement aux Principes des Connoissances", *Neue Leipziger Literaturzeitung,* 117, 1804, col. 1860–1872; Review of "Histoire comparée des Systèmes de Philosophie considérés relativement aux Principes des Connoissances", *Jenaische Allgemeine Literatur-Zeitung,* 5, 1808, col. 33–40; *Jenaische Allgemeine Literatur-Zeitung,* 6, 1808, col. 41–48; *Jenaische Allgemeine Literatur-Zeitung,* 7, 1808, col. 49–51. The English and American reception also shows that his history was widely read. See G. H. Lewes, *A Biographical History of Philosophy, Series I. Ancient Philosophy, Vol. 1* (London: Charles Knight, 1845), p. 10: "This work enjoys considerable reputation, and deserves it"; cf. also the review in the *London Critical Review:* Review of "Histoire comparée des Systèmes de Philosophie considérés relativement aux Principes des Connoissances", in *Critial Review* (London: S. Hamilton, 1804), pp. 559–571. For the American reception cf. Review of "Histoire comparée des Systèmes de Philosophie considérés relativement aux Principes des Connoissances", *North American Review,* 43 (1824): 234–266 (p. 235); Gérando is praised here for not having adopted the Germans' "peculiar modes of thinking".
17. Cf. Thierry Hoquet, "History without Time: Buffon's Natural History as a Nonmathematical Physique", *Isis,* 101/1 (2010): 30–61.
18. Joseph Marie de Gérando, *Histoire comparée des systèmes de philosophie considérés relativement aus principes des connaissances humaines*, v. 1 (Paris: Alexis Eymery, 1822 [1804]), p. xiii.
19. *Ibid*, xiv.
20. *Ibid*, xvii.
21. *Ibid*, xxiv–xxv. This idea is also present in the *Leçons d'histoire* (Lessons of history) given by Constantin-François Volney at the *École normale* in 1795. Constantin-François Volney, *Leçons d'histoire prononcées a l'École normale, en l'an III de la République Française* (Paris: Chez J.A. Brosson, 1800), pp. 158–188.
22. Joseph Marie de Gérando, *Histoire comparée des systèmes de philosophie considérés relativement aus principes des connaissances humaines*, v. 1, p. xxvii: "être considéré comme l'essai d'un traité de philosophie, sur les questions premières de cette science, mais d'un traité conçu d'après la méthode la plus prudente, quoique la plus négligée, d'après la méthode des expériences; nous osons donc l'offrir encore comme une introduction à l'histoire de la philosophie ou comme un essai de philosophie expérimentale".
23. In the second half of the eighteenth century there were many attempts in Germany to define the concept of history of philosophy. For an overview, see Georg Friedrich Daniel Goess, "Über den Begriff der Geschichte der

Philosophie", in *Über den Begriff der Geschichte der Philosophie und über das System des Thales. Zwo philosophische Abhandlungen*, ed. Georg Friedrich Daniel Goess (Erlangen: Iohann Iakob Palm, 1794), pp. 7–50. It was Victor Cousin, a friend of Gérando's, who criticized his approach to writing history. He emphasized that Gérando was too much orientated towards his method instead of narrating past events as they had really happened. Victor Cousin, Review of "Histoire comparée des Systèmes de Philosophie, par M. Degérando: deuxième édition, revue, corrigée et augmentée", *Journal des savants* (July, 1825): 434–439 (p. 437).

24. Among other works of the time now forgotten philosophers of history like Karl Leonhard Reinhold and Georg Gustav Fülleborn offered the theoretical framework which Gérando cited as a reference for his idea of history as philosophical method of observation. Cf. Karl Leonard Reinhold, "Über den Begriff der Geschichte der Philosophie", in *Beiträge zur Geschichte der Philosophie*, ed. Georg Gustav Fülleborn (Züllichau-Freystadt: Frommann, 1791), p. 32.
25. Gérando affirmed that the *observateur* of history had to be at an appropriate distance to his facts and commit to "constant neutrality". See Joseph Marie de Gérando, *Histoire comparée des systèmes de philosophie considérés relativement aux principes des connaissances humaines*, v. 1 (Paris: Eymery, 1822), p. xxxvii, p. 55.
26. *Ibid*, v. 1, p. xvii f.
27. Joseph-Marie de Gérando, *Des signes et de l'art de penser considérés dans leur rapports mutuels*, v. 4 (Paris: Chez Goujon, 1799), p. 561: "toutes ces réformes réunies ne feroient pas disparoitre entierement de nos langues les divers défauts que nous y sentons, ne satisferoeint pas aux conditions du langage parfaitement philosophique, et que par la nature même des choses plusieurs de ces conditions sont incompatibles les unes avec les autres, du moins dans leurs accomplissement".
28. *Ibid*, 570–573.
29. For more on the concept of communication within Gérando's thought, see his editorial article "Des communications littéraires et philosophique entre les nations de l'Europe", Archives littéraire de l'Europe ou Mélanges de littérature d'Histoire et de Philosophie, 1 (1804): 1–18.
30. Joseph-Marie de Gérando, "Considérations sur les diverses méthodes à suivre dans l'observation des peuples sauvages", *Revue d'anthropologie*, 12 (1883 [1800]), pp. 153–182 (p. 155).
31. In fact, for Gérando, the first task of the voyager was to learn the savage's language. *Ibid*, 159.
32. *Ibid*, 155.
33. Sven Ködel, *L'Enquête Coquebert de Montbret (1806–1812) sur les langues et dialectes de France et la représentation de l'espace linguistique français sous le Premier Empire*. Unpublished Doctoral Dissertation, University Paris 7, 2013.
34. "Ce n'est point sous le rapport de la necessité d'etre entendue, que j'ai soumis à *Votre Excellence*, l'idée de rendre à la Langue Catalane le droit de servir aux actes publics. Je n'ai présenté cette idée que sous des rapports politiques; j'ai cru d'avoir faire remarquer que l'usage de la langue Castillane ou espagnole n'avait été introduit en Catalogne qu'à l'époque de la submission de cette province aux loix generales de l'espagne dont elle s'etoit conservée jusq'alors indépendante . . . Le Catalan hait l'espagnol, il ne veut pas être espagnol . . . on pourroit se prévaloir de souvenir du passé, des antipathies existantes, pour commander plus facilement en l'opinion, et le choix de la langue autorisée dans les actes publics est peut être une des choses qui influe le plus à cet égard sur l'opinion nationale". Archives Nationales (France) AN BB/ 5/ 287—Ordre Judiciaire Catalonie. For an account of the linguistic policies under the

Napoleonic rule in Catalonia and a reproduction of Gérando's letter, see Rolf Kailuweit, "Sprechen und Schweigen: das Scheitern der französischen Sprachpolitik im besetzten Katalonien 1810", in *Polyglotte Romania. Homentage a Tilbert Stegman*, vol. 1, eds. Brigitte Schlieben-Lange and Axel Schönberger (Frankfurt am Main: Domus Editoria Europaea, 1991), pp. 295–337.

35. Letter to the Baronne Fritz de Dietrich, January 1802. Marie Anne de Gérando, *Lettres de la baronne De Gérando née a Rathsamhausen suivies de fragments d'un journal écrit par elle de 1800 a 1804* (Paris: Librairie académique, 1880), p. 180.
36. Nathalie Vuillemin, "Connaissance ou reconnaissance? L'inquiétante corruption du regard par le savoir dans l'histoire naturelle des Lumières", *Variations*, 12 (2004), pp. 57–73 (p. 68).
37. Sicard, *Cours d'instruction*, p. xlv.
38. Charles Maurice de Talleyrand-Périgord, *Rapport sur l'instruction publique* (Paris: Imprimerie de Baudouin, 1791), p. 100: "Il faut . . . que les anciennes formes obséquieuses, ces précautions timides de la foiblesse, souplesses d'un langage détourné qui sembloit craindre que la vérité ne se montrât toute entière . . . se perde dans un langage simple, fier et rapide; car là où la pensée est libre, la langue doit devenir prompte et franche".
39. *Ibid*: "qui est l'image vivante de la pensée, . . . par qui les idées même les plus abstraites deviennent presque visibles, et qui, dans sa décomposition, . . . présente la véritable grammaire, non des mots, mais des idées".
40. Pierre-Louis Roederer, "Review of Sicard, Élémens de grammaire générale appliqués a la Langue Française", in *Oeuvres de Come P.L. Roederer. Publiées par son fils* (Paris: Typographie de Firmin Didot Frères, 1859 [1799]), pp. 28–29 (p. 29): "Cet ouvrage serait d'une suprême utilité, non-seulement pour les sourds-meuts, mais pour l'humanité toute entière . . . On y trouvera probablement un principe nouveau d'une pasigraphie parfaite, et peut-être même d'une langue philosophique et universelle".
41. Roch Ambroise Cocurron Sicard, *Élémens de grammaire générale, appliqués a la langue Francaise*, v. 1 (Paris: Deterville, 1801), p. xiv: "On sait bien que les premières langues ont été le produit du hasard auquel ne présida aucun genre d'analise [sic]".
42. Sicard, *Cours d'instruction*, p. xiv: "Il faut donner un nouvel être à cet automate qui n'est qu'animé, établir quelque lien de communication entre lui et les autres hommes".
43. *Ibid*, xlv: "institution d'un sauvage mise en action, dans laquelle l'instituteur, profitant du très-petit nombre d'élémens connus de la grammaire de cet homme de la nature, créera avec lui . . . la grammaire de l'homme de la société".
44. Roch Ambroise Cocurron Sicard, *Cours d'instruction d'un sourd-muet de naissance* (Paris: Le Clere, 1799), p. 501: "une connoissance parfaite des moindres opérations de l'entendement; . . . une patience d'observation, qui, à l'exemple de l'Anatomiste, qui tient compte de la fibre la plus déliée".
45. Roch Ambroise Coccuron Sicard, *Élémens de grammaire générale appliqués à la langue française*, v. 1, p. 11.
46. *Ibid*, v. 1, xvii–xviii.
47. Jean Massieu (together with Laurent Clerc) was one of the most famous pupils of Sicard. Sicard knew him before he got his post at the deaf-mute's school in Paris while he was teaching at Bordeaux. He travelled with Massieu and Clerc to London where he tried to give the same public lessons as in Paris. Thus, they were crucial for Sicard to gain scientific credit in public. See Harlan Lane, *When the Mind Hears* (New York: Random House, 1985).
48. See Karl Asmund Rudolphi, *Bemerkungen aus dem Gebiet der Naturgeschichte, Medicin und Thierarzneykunde, auf einer Reise durch einen Theil*

von Deutschland, Holland und Frankreich (Berlin: Realschulbuchhandlung, 1804), pp. 288–291, and Anonymous, "Ueber das Taubstummen-Institut zu Paris im Januar 1805", *Isis*, 4 (1806): 209–223 (p. 223).

49. See August von Kotzebue, *Erinnerungen aus Paris im Jahre 1804* (Carlsruhe, 1804), pp. 154–159.
50. Sophia Rosenfeld, "Deaf Men on Trial: Language and Deviancy in Late Eighteenth-Century France", *Eighteenth-Century Life*, 21.2 (1997): 157–175.
51. Roch Ambroise Sicard, *Cours d'instruction d'un sourd-muet de naissance, pour servir a l'éducation des sourds-muets* (Paris: Le Cere, 1799), p. xxxi: "tout langage, dit notre philosophe, n'est qu'une collection des signes, comme une suite de dessins d'histoire naturelle est une collection d'images, une représentation d'un grand nombre d'objets. On peut tout figurer par gestes, comme on peint tout par des couleurs".
52. Louis-François Jauffret, "Introduction aus mémoires de la société des observateurs de l'homme", in *Aux origines de l'anthropologie française. Les mémoires de la Société des observateurs de l'homme en l'an VIII*, eds. Jean Copans and Jean Jamin (Paris: Le Sycomore, 1978), pp. 71–87 (p. 81).
53. Roch Ambroise Sicard, *Cours d'instruction d'un sourd-muet de naissance, pour servir a l'éducation des sourds-muets* (Paris: Le Cere, 1799), p. vi.
54. Johann Conrad Amman, *The Talking Deaf Man: Or, a Method Proposed, Whereby He Who Is Born Deaf, May Learn to Speak* (London: Printed for Thomas Howkins, 1694), p. 6. For the dispute between the advocates of an "oral" approach and of the "sign language" approach, see François Buton, "L'éducation des sourds-muets au XIXe siècle. Description d'une activité sociale", *La découverte*, 223 (2008): 69–82.
55. Sicard, *Grammaire générale*, v. 1, p. 15.
56. François Buton "Ce qu'administrer veut dire. Gérando et l'education des sourds-muets", in Chappey, Christen, and Moullier, *Joseph-Marie de Gérando*, pp. 143–156.
57. Antoine-Isaac Sylvestre de Sacy, Review of "De l'Éducation des Sourds-Muets de Naissance", *Journal des Savans* (1828): 323–338 (p. 324): "Le premier s'appliquera à en approfondir la théorie, le second à recueillir les faits, et le troisième, à chercher dans les leçons de la théorie et dans les résultats de l'expérience, le moyen d'éviter les écueils, de remplir les lacunes, d'assurer les succès, et de simplifier et régulariser les méthodes".
58. Charles-Maurice de Talleyrand-Périgord, *Proposition faite à l'Assemblée nationale sur les poids et mesures* (Paris: Imprimierie Nationale, 1790), pp. 10–11. See also Ken Alder, *The Measure of All Things: The Seven-Year Odyssey and Hidden Error that Transformed the World* (New York: Free Press, 2002); Charles Gillispie, *Science and Polity in France: The Revolutionary and Napoleonic Years* (Princeton, Oxford: Princeton University Press, 2004), pp. 227–447.
59. The Abbé Grégoire, as Patrice Higonnet relates, estimated in 1794 that French was the dominant language only in 15 of the 89 départements, that 6 million French people could not speak French at all, and that 6 million could not properly speak French (there were 30 kinds of *patois* which were spoken at that time). Patrice Higonnet, "The Politics of Linguistic Terrorism and Grammatical Hegemony during the French Revolution", *Social History*, 5.1 (1980): 41–69 (p. 49). See also the report Grégoire made before the *Institut national* with the title "sur la nécessité d'anéantir les patois et d'universaliser l'usage de la langue française", Anne and Henry Denys, "Six Lettres inédites de Volney a Grégoire", *Dix-Huitième siècle*, 23 (1991): 233–245 (p. 235).
60. Roch Ambroise Sicard, *Cours d'instruction d'un sourd-muet de naissance, pour servir a l'éducation des sourds-muets* (Paris: Le Cere, 1799), pp. xlvii–xlviii:

"une espèce de langue universelle qui . . . peut être entendue par les hommes de tous les pays et de toutes les nations".

61. Thus Gérando states in his Report to the Emperor of 1808: "Amie des lois, [la Philosophie] leur assure une obéissance éclairée et raisonnable, la seule qui soit digne des bonnes lois ; elle enseigne le respect à l'ordre établi, rappelle les hommes aux instructions de l'expérience, et repousse les innovations téméraires" (Friend of the laws, philosophy guarantees an enlightened and reasonable obedience, the only worthy of good laws; she teaches respect for the established order, reminds men of the instructions of experience, and pushes back daring innovations). Joseph Marie de Gérando, "Philosophie", in *Rapport Historique sur les Progrès de l'Histoire et de la littérature ancinnce depuis 1789 et sur leur état actuel. Présenté à Sa Majesté l'Empereur et Roi, en son Conseil d'état, le 20 Février 1808, par la Classe d'Histoire et de la Littérature ancienne de l'Institut,* ed. Bon-Joseph Dacier (Paris: Imprimerie Impériale, 1810), pp. 278–355 (pp. 280–281).
62. Joseph Marie de Gérando, *Rapport fait à la société de Paris pour l'instruction élémentaire, À l'Assemblée Générale du 19 février 1816* (Paris: Colas, 1816), p. 5.

6 Contested Boundaries

How Scientists Deal with Uncertainty and Ambiguity in Language

Priya Venkatesan Hays

1. INTRODUCTION

Language is a tool for scientific production. Language allows scientists to get their message across and communicate their findings, and it plays a key role in the process of building scientific arguments and shaping scientists' ideas. Language is not only a tool of science but serves as a context in which science develops.[1] However, language in its malleability and fluidity sometimes manages to get out of control from the hands of scientists. This article is a study of how language escapes the rigidity of scientific work and serves as a means by which science expresses doubt, ambiguity, and uncertainty. Interestingly, while many politicians have embraced this doubt, particularly in debates about climate change, scientists, in positivist fashion, have engaged in meaningful efforts to erase ambiguity and uncertainty in their work. In fact, on the question of uncertainty, ambiguity, and doubt in scientific language, politicians and scientists seem to be on opposite ends of the epistemological spectrum. Politicians would like for scientists to caveat their claims. For example, John Beddington, the UK government's chief scientific advisor from January 1, 2008, to the present, thinks the key challenge in the global discussion of climate change is that scientists do not caveat their work extensively enough. Beddington stated that "certain unqualified statements have been unfortunate. We have a problem in communicating uncertainty . . . All of these predictions have to be caveated by saying, 'There's a level of uncertainty about that' ".[2]

However, among scientists the opposite trend is occurring. Scientists are proposing the elimination of uncertainty and ambiguity in their output. Yet many studies claim that due to the nature of scientific methodology, research articles are rich in speculative and tentative statements, also known as hedges, and that to counter this, a linguistically motivated approach is used to address the problem of recognizing such language in biomedical research articles.[3] Other studies claim that biomedical documents contain a range of terms with more than one possible meaning and that these ambiguities form a significant obstacle to the processing of biomedical texts.[4] Databases such as BioScope are being produced to handle negation and uncertainty in biomedical texts.[5]

This paper addresses how scientists mediate the uncertainty and ambiguity of language to manage the rhetorical effects of language in their research articles. Scientists seem to have disavowed the embedding of certain qualities inherent in language in their texts, which politicians have unwittingly espoused for political reasons. This form of hedging is also denoted as powerless language, or language that is uncertain or heavily qualified, as opposed to powerful language, that is, language that is certain and assertive.

Uncertainty and ambiguity serve as an element of medical and scientific discourse, particularly surrounding areas of controversy, and scientists have devised evidentiary ways of dealing with it. According to Kinzig and Starrett:

> Throughout their history, humans have devised a number approaches and strategies to reducing (perceived) uncertainty or the psychological impacts of uncertainty, and to function and thrive in a partially unknown world. A certain world begets certain "best strategies" for survival; an uncertain world necessitates a larger portfolio of options and approaches, and means for choosing among them . . . The scientific process is built on the goal of advancing knowledge; penetrating and reducing the reaches of what is not known. Each advance is built on knowledge acquired earlier. The cost of "incorrect knowledge" is therefore quite high, affecting not only that building block in the foundation but those that follow. Science has therefore evolved procedures whose primary goals are to protect against being wrong . . . Scientists apply relatively high evidentiary standards—insisting, for instance, that there be only a small probability (sometimes 5% but frequently lower than 1%) that an incorrect conclusion has been drawn. This means that many conclusions concerning the workings of the world that have a reasonable probability of being correct—say 75%—may not pass muster.[6]

In other words, scientists withstand a long-held view that their adoption of language leads ultimately to the charge of anti-realism by holding to the highest standards of probabilities in making their conclusions.[7] Efficiency is put at a premium over generalizability. As Susan Leigh Star writes:

> Scientists constantly face uncertainty. Their experimental materials are recalcitrant; their organizational politics precarious; they may not know whether a given technique was correctly applied or interpreted; they must often rely on observations made in haste or by unskilled assistants. As many observers of science have noted, these contingencies rarely appear in published descriptions of scientific work.[8]

Referring to the anthropological studies of Karin Knorr-Cetina and Bruno Latour and Steve Woolgar,[9] the sociological work of Michael Lynch,[10] and the articles on semiotics of science by Francoise Bastide,[11] Leigh Star also notes:

> Knorr-Cetina's and Latour and Bastide's work on the transmogrification and deletion of uncertainties from lab work to published reports have emphasized this point; similarly, Latour and Woolgar's *Laboratory Life*

> documents in some detail the transformation of everyday uncertainties into facts via "deletion of modalities" and progressive reification. Lynch's work on artifacts and on neurobehavioural tests notes similar phenomena under two different types of conditions. The work of Woolgar and that of [Trevor] Pinch has also described a similar situation. One major thrust of this research thus far has been to document the presentation of data, from observation to publication, as increasingly certain. The published data reveal, rather than hint; articles state, rather than guess at; subjects line up and are counted—by the time they get to the journals, they don't run away and hide behind the lab equipment or try sabotaging experiments. These observations about the deletion of uncertainty have added valuable insight about the process of conducting science.[12]

Doing away with uncertainty is part of the process of performing science. Social studies of science, through the work of Knorr-Cetina, Latour, and Lynch, have shown that the contexts of science, particularly the social ones, imbue the life of the laboratory with a complexity unbeknownst to the reader of a journal article. On the surface, the completed production of scientific text understates the social processes that accompanied the technical work, social processes that are part and parcel of generating uncertainty. Sociologists of science demonstrate that uncertainty is part of scientific production, underscoring arguments about the validity of scientific claims. Unwittingly, perhaps, they have made strange bedfellows with politicians who call for reservations to be made about scientific claims.

Whether this proclivity towards scientific certainty leads to greater credibility has been a source of controversy. Jensen performed a multiple-messaging experiment in which news stories of cancer research were read by college students. The credibility of these news stories was then assessed on numerous levels. Jensen writes of the results:

> News reports of scientific research are rarely hedged; in other words, the reports do not contain caveats, limitations, or other indicators of scientific uncertainty. Some have suggested that hedging may influence news consumers' perceptions of scientists' and journalists' credibility (perceptions that may be related to support for scientific research and/or adoption of scientific recommendations). But whether hedging does affect audience perceptions is unknown. A multiple-message experiment (N = 601) found that across five messages, both scientists and journalists were viewed as more trustworthy (a) when news coverage of cancer research was hedged (e.g., study limitations were reported) and (b) when the hedging was attributed to the scientists responsible for the research (as opposed to scientists unaffiliated with the research).[13]

In this context, I will examine two such areas of controversy in science where the language of uncertainty enters into scientific frameworks: climate change and human embryo research. These controversies are undoubtedly part of the discourse of public institutions, government, education, and political life in general.

1. UNCERTAINTY IN SCIENTIFIC DISCOURSE: THE CASE OF CLIMATE CHANGE

Stephen Zehr, who was former program director for the Science, Technology, and Society Program at the National Science Foundation in the U.S., conducted a study of uncertainty in language surrounding journalistic accounts of global warming framed in scientific discourse. His article in *Public Understanding of Science* argues that journalistic accounts of science serve as a key indicator of politicians' and scientists' comparable and contrasting views on scientific issues and show the degree to which each have input in the overall framing of scientific issues to the public.[14] Zehr reiterates Leigh Star in stating:

> Constructivist studies of scientific knowledge have emphasized how uncertainties in knowledge production processes are eliminated or transformed, resulting in more certain claims about the natural world. Recent research also has problematized scientific claims about uncertainty and ignorance in science.[15]

Within uncertainty discourse, Zehr includes the individual or collective claims that scientific knowledge is incomplete about some feature of nature. These claims can be explicit, such as in the statement, "There is scientific uncertainty about the causal linkage between twentieth-century greenhouse gas emissions and global temperature increases".[16] Zehr characterizes an uncertainty discourse as one in which scientists can make contrasting claims or inferences about nature or use uncertainty as a means of characterizing opposing scientific claims. Zehr concurs with many sociologists of science in stating that modalities, or qualifying statements, are employed by scientists to mitigate uncertainty in their work.

However, according to Zehr, less attention has been placed on the implications of scientific uncertainty for the relationship between scientists and the public. Representations of uncertainty could potentially diminish the authority of science. Since science is generally perceived as an authoritative, truth-producing power, acknowledgment of uncertainty reduces this authority and opens access to the production of knowledge to other groups and activities. On the other hand, representations of uncertainty might be managed to augment scientific authority if they place the problem in question more firmly in future scientific research. In other words, there is a dual role for uncertainty in science. In the case study reported here, Zehr suggests that the discourse of uncertainty in climate change controversies actually placed it within future scientific research. Uncertainty discourse was used in the press to situate science more centrally within global warming/climate change and future studies on it. Science's footing was re-established by charges of uncertainty, leading to calls for more research (by scientists) to be conducted. This suggests that the discourse of uncertainty may be a more benign trait in scientific discussions and actually could lead to increasing science's status and authority, rather than further problematizing scientific domains.

The salience of scientific uncertainty was reflected in several features of the following articles on global warming and climate change in the popular press. In many of these journalistic accounts, scientific uncertainty was the main topic. For example, a high-profile 1991 *New York Times Magazine* article entitled "Heating the Global Warming Debate" focused extensively on scientific uncertainty in the debate over rates of global warming.[17] K. Wright situated the problem with an initial large print statement that read: "In 1988 scientist Jim Hansen testified that the world was getting hotter: But how hot? And how fast?"[18] Several other articles also situated scientific uncertainty as the main topic. In other articles, scientific uncertainty appeared in key places. These included titles such as: "Cloud Data Cast Doubt About Atmosphere";[19] "Global Warming: Experts Ponder Bewildering Feedback Effects";[20] "With Cloudy Crystal Balls, Scientists Race to Assess Global Warming";[21] "Global Warming: Uncertainty and Action";[22] and "U.S. Water Resources versus an Announced But Uncertain Climate Change".[23] These articles included an opening or closing paragraph (or both) that helped frame uncertainty within the article. A 1991 *New York Times* article on an international global warming meeting had a typical beginning: "In a contest between uncertain science and uncertain economics, negotiators from around the world convened in Nairobi yesterday for what promises to be a contentious effort".[24] Further references to uncertainty were placed at several points throughout the article.[25]

Other examples abound of uncertainty discourse appearing in journalistic framing of climate change, questioning scientific claims about increasing CO^2 levels in the atmosphere, rising water levels, and melting polar caps. According to Zehr, journalists also framed scientific discourse as having aspects of uncertainty rather than directly quoting scientists. In addition, even though "direct quotations from scientists hold more force than narrative, there were many instances in which these quotations included uncertainty discourse".[26] Adler and Hadorn discussed the treatment of uncertainties in climate change discourse within the Intergovernmental Panel on Climate Change (IPCC) assessment. In their review of the literature, where they analyzed articles such as "Reasoning about Climate Uncertainty",[27] "Improving Conveyance of Uncertainties in the Findings of the IPCC",[28] and "Certainty, Uncertainty, and Climate Change",[29] they concluded with concerns about if "the guidance note is adequate for the purpose of communicating clear and relevant information to users" and statements that uncertainty is a source of relevant information for users (policymakers).[30] This type of uncertainty in language has been espoused by a number of politicians, including Beddington and members of the U.S. Congress who wish to caveat claims about global warming and climate change among scientists themselves. Journalists also adopt this uncertainty discourse.

2. UNCERTAINTY IN CLAIMS ABOUT HUMAN EMBRYO RESEARCH

The language of uncertainty also appears in scholarly discourse about the human embryo. Discussion of human embryo research has typically taken place in the realm of bioethics. In his article "Human Embryo Research and Moral Uncertainty", William Cheshire argues that the language reflected on the embryo is framed by uncertainty and that this uncertainty serves as a marker of each author's particular ethical standpoint on human embryo research. His study found, within the text of 53 articles[31] reporting on human embryo research, 139 non-neutral words or phrases relevant to the evaluation of moral status of the human embryo. Specific word choices reflected a range of judgments on the questions of whether the human embryo is a form of life, whether the human embryo is a human life, and what level of moral worth the embryo, as human, either has intrinsically or is given. Designations relating to the degree or category of humanness varied considerably. Thirteen articles explicitly referred to the human embryo as a form of "human life", and three others as a "human being", whereas nine articles explicitly claimed or quoted experts as claiming that the human embryo is not a human life. Two others explained that the embryo is not an "entire human being", another two that the embryo is not an "actual" human being, and one that the embryo is not an "existing human". One article classified the embryo as a "potential human life". Another stated that the embryo is not a "human individual". Other articles distinguished human embryos as belonging to a category separate from babies or children, one claimed embryos were distinct from fetuses, and two claimed embryos were distinct from adults. Four articles used the word "respect" as an attitude that might apply to human embryos. Three articles used the word "kill" in connection with the procedure that extracts stem cells from embryos, two others used "death", one "dismembering", and one "murder". Two articles referred to the notion that human embryos might have civil "rights".[32] No articles referenced the genetic completeness or uniqueness of the embryo but focused rather on the lack of development. Research articles referred to the embryos as "early stage", a "cluster of cells", or "just a clump of undifferentiated cells". One article stated that cells are progenitors of "blank slates".[33]

These case studies of climate change and embryo research and the uncertainty inherent in the discourse surrounding them point to the fluidity and multiplicity of language, language that in its permissibility allows for its use as a rhetorical tool that can engender multiple interpretations. Hesitancy, ambiguity, caveating, argumentation, and hedging are highlights of this type of uncertainty in discourse leading one to believe that science may partly belong in the category of rhetoric.

3. HOW SCIENTISTS MANAGE UNCERTAINTY AND AMBIGUITY IN LANGUAGE

It is this "uncertain" aspect of the commingling of science and language, expressing doubt and ambiguity, which scientists have developed linguistic tools to mitigate. In fact, the opposite trend is occurring among the enclaves of scientists, where hesitancy, speculation, and uncertainty are objects of linguistic elimination. From databases to datasets to weighting and hedging, assertion or factual language is the objective in formal scientific language. In a survey of work on how scientists react to uncertainty and ambiguity in language, though speculation and ambiguity may be normative, they are undesirable, and attempts to reposition them may detract from the richness of their use of language.

As Kilicoglu and Bergler begin their article "Recognizing Speculative Language in Biomedical Research Articles: A Linguistically Motivated Perspective",[34] in recognizing speculative language in biomedical research articles, the scientific method involves making hypotheses, experimenting, and reasoning to reach conclusions, which are often tentative and provisional. Scientific writing, particularly in biomedical research articles, reflects this, as it is rich in speculative statements (hedges). Most text processing systems ignore hedging and focus on factual language (assertions). Although assertions, or even mere co-occurrence of terms, may be sufficient for some information extraction and text mining applications, identifying hedged text is crucial because hedging acts as a valence shifter, altering, and in some cases even reversing, factual statements. For instance, the italicized fragment in example (1a) implies a factual statement while example (1b) is considered speculative:

(1a) Each empty cell indicates that the corresponding TPase query was not used at the particular stage of PSIBLAST analysis.

(1b) The lack of Cut expression in wild-type ventral cells abutting the D-V boundary indicates that D-mib is required for Ser signaling by dorsal cells and acts in a non-autonomous manner to activate N in ventral cells.[35]

These examples not only illustrate the phenomenon of hedging in the biomedical literature, they also highlight difficulties in recognizing hedges. The word *indicate* plays a different role in each example, acting as a hedging cue only in (1b). Hedging in the second sentence seems to be further marked by the subject of the verb *indicate*: "The lack of Cut expression in wild-type ventral cells abutting the D-V boundary".

The term *hedging* was introduced into the linguistic literature by Lakoff.[36] He proposed that natural language sentences can be true or false *to some extent*, contrary to the dominant truth-conditional semantics paradigm of the era. He was mainly concerned with how words and phrases, such as *mainly* and *rather*, make sentences fuzzier or less fuzzy. Palmer identifies epistemic modality, which expresses the speaker's degree of commitment to the truth of propositions and is closely linked to hedging.[37] He identifies three types of epistemic modality: speculatives to express uncertainty, deductives to indicate an inference from observable evidence, and assumptives to indicate inference from what is generally known. He focuses mainly on the use of modal verbs in expressing these types. Chafe uses the term *evidentiality* to describe the hedging phenomenon while adopting a narrower definition of hedges, referring only to the linguistic devices that indicate a less than perfect match between a piece of knowledge and a category, such as *about* and *sort of*.[38] Hyland provides a comprehensive account of hedging in scientific articles from a linguistic perspective.[39] He views hedges as polypragmatic devices with an array of purposes (e.g., weakening the force of statement, expressing deference to the reader, signaling uncertainty) and proposes a fuzzy model in which he categorizes scientific hedges by their pragmatic purpose, such as content-oriented hedges, writer-oriented hedges, and reader-oriented hedges. He also identifies the principal syntactic realization devices for different types of hedges, including epistemic verbs (verbs indicating the speaker's mode of knowing), adverbs, and modal auxiliaries, and presents the most frequently used members of these types based on analysis of a corpus of molecular biology articles. He finds that hedges are similarly distributed between abstracts and full-text and that they are most frequent in Results and Discussion sections of scientific articles. Kilicoglu and Bergler's results from this article demonstrate that speculative language can be recognized successfully with a linguistically motivated approach that confirms that selection of hedging devices affects the speculative strength of the sentence, which can be captured reasonably by weighting the hedging cues.[40]

Stevenson et al. address lexical ambiguity, the linguistic phenomenon where a term has more than one potential meaning.[41] They employ WSD, word sense disambiguation, to resolve these lexical ambiguities. Some words have multiple senses; for example, the word *culture* could mean an anthropological culture, such as the culture of a Mayan civilization, or a laboratory culture, such as a cell culture. The different senses of a word are often obtained from a sense inventory such as a dictionary or other resource. The Unified Medical Language System (UMLS) is one such sense inventory for the biomedical and clinical domain. In the UMLS, senses (or concepts) associated with words and terms are enumerated via Concept Unique Identifiers (CUIs). For example, the two senses of culture are "C0010453: Anthropological Culture" and "C0430400: Laboratory Culture" in the UMLS release 2007AB. The UMLS is more than just a dictionary of different word senses but also a framework encoded with different semantic and syntactic structures. Some such information includes related concepts, semantic types, and semantic relations. A semantic type is a broad subject categorization assigned to a CUI. A semantic relation is the relationship between two semantic types. For example, the semantic type of "C0010453: Anthropological Culture" is "Idea or Concept" while the semantic type for "C0430400: Laboratory Culture" is "Laboratory Procedure". The semantic relations between "Idea or Concept" and "Laboratory Procedure" with the semantic type "Mental Process" are "result of" and "assesses_effect_of", respectively.

MetaMap 1 maps terms in biomedical texts to senses (i.e., concepts) in the UMLS by identifying the CUIs of the content words in the text. MetaMap can be thought of as an all-words disambiguation system. Another approach is focused on particular target words. MetaMap assigns a CUI (sense) to every word or term that it can in a running text using rules and patterns. This approach is based on supervised learning, where we collect some number of manually disambiguated examples of a given word and derive a model from that data that only assigns senses to that target word. Thus, MetaMap is a broad coverage tool while the latter approach is more fine-grained and specific to a few words.

Detecting uncertain and negative assertions is essential in most biomedical text mining tasks where, in general, the aim is to derive factual knowledge from textual data. Vincze et al. report on a corpus annotation project that has produced a freely available resource for research on handling negation and uncertainty in biomedical texts (they call this corpus the BioScope corpus).[42] The corpus consists of three parts, namely medical free texts, biological full papers, and biological scientific abstracts. The dataset contains annotations at the token level for negative and speculative keywords and at the sentence level for their linguistic scope. The annotation process was carried out by two independent linguist annotators and a chief linguist—also responsible for setting up the annotation guidelines—who resolved cases where the annotators disagreed. The resulting corpus consists of more than 20,000 sentences that were considered for annotation, and over 10% of them actually contain one (or more) linguistic annotation suggesting negation or uncertainty. One such annotation is the following sentence, which can describe two different situations: *There is [no] primary impairment of glucocorticoid metabolism in the asthmatics.* First, the glucocorticoid metabolism is impaired in the asthmatics, but not primarily, that is, the scope of "no" extends to "primary" only: ([no] primary). Second, the scope of "no" extends to "impairment" (and its modifiers and complements as well); thus, there is no impairment of the glucocorticoid metabolism at all: ([no] primary impairment of glucocorticoid metabolism in the asthmatics). Another example is *Mild viral <or> reactive airways disease is detected.* The syntactic structure of this sentence is ambiguous. First, the airways disease is surely mild, but it is not known whether it is viral or reactive: (viral <or> reactive); or second, the airways disease is either mild and viral or reactive and not mild (mild viral <or> reactive). Most of the sentences with similar problems cannot be disambiguated on the basis of contextual information; hence the proper treatment of such sentences remains problematic.

These methods outline the ways in which scientists deal with uncertainty in language. Uncertainty is a marker for hedging and speculation, which deters the textual mining and automatic processing of biomedical texts and thus being the focus of scientists' seemingly unproblematized strategies. While certain politicians would submit to uncertainty in language, scientists on the other hand share a conviction in the need to categorize and eliminate uncertainty in language. Thus uncertainty is a contested boundary, wherein politicians, journalists, scholars, and scientists confront uncertainty, ambiguity, and doubt, albeit with different methods and purposes.

4. IMPLICATIONS FOR THE CORRESPONDENCE THEORY OF TRUTH

According to the philosopher of science and scientist Alfred Tauber, in any discussion of language and science, "no less than reality is at stake".[43] In evaluating the work of W.V.O. Quine, Rudolf Carnap, and Otto Neurath, twentieth-century philosophers of science who explored how the use of language by scientists pre-determines whether science can in fact mirror nature, Tauber notes that "in recognizing that our conceptions of reality are determined by language, [language] cannot be escaped nor viewed from afar".[44] According to Quine:

> The fundamental-seeming philosophical question, How much of science is merely contributed by language and how much is a genuine reflection of reality? Is perhaps a spurious question which itself arises wholly form a certain particular type of language. Certainly we are in a predicament if we try to answer the question; for to answer the question we must talk about the world as well as about language, and to talk about the world we must already impose upon the world some conceptual scheme peculiar to our own special language.[45]

Indeed, from this analysis, Tauber concludes that "'correspondence' theories of truth are therefore forbidden".[46] The ethical implications of the analysis presented here are somewhat consistent with Tauber's vision in that science may have a circumspect relation to reality through its use of language. Society may in fact see science as tantamount to truth, while the contested boundaries surrounding the politics of language in science undermines this perception to some degree. Scientists are manipulating language for certain ends to reflect or imply correspondence. While philosophically this would imply that the correspondence theory of truth is suspect, socially the status of science may be in question. Ethically, science's access to truth is also under scrutiny. This occurs as a result of science's relation to the politics. The political milieu of issues such as global warming and embryonic research allows for conflicting scientific viewpoints as a result of political will.

As mentioned earlier, the likes of philosophers of science such as Tauber make strange bedfellows with conservative politicians who question scientific processes' access to truth, particularly concerning climate change. Most scientists claim to eliminate uncertainty and ambiguity from their frameworks, and they do so vividly with rigor, as suggested here. The status of science is not in question for these researchers. The question remains whether society should consider the uncertainty and ambiguity resulting from scientific uses of language or see it as a consequence of muddying politics. This question is at the heart of science's confrontation with language. The answer to this question may lie in a study of science from an ethical perspective, which reconciles the clarity required of language in science with the political uses of science.

However, according to Harmon and Gross, discussion sections of scientific papers *ought to* reflect hedging lest the claims of the scientists seem misleading:

> The discussion section ought normally reflect what would in real life be an abnormal condition: a split personality. One personality boldly makes claims about the nature and causal structure of the material world; the other tempers those claims in an attempt to avoid misleading the community of scientists . . . In the discussion, then, scientists routinely interpret their results so as to make the broadest claims they plausibly can, given their evidential base. Nevertheless, they must at the same time be extremely cautious about making broad claims, lest their work be dismissed as rank speculation.[47]

Language in science is a fluid entity whereby scientists try to eliminate doubt and ambiguity, whereas political viewpoints aim to undermine these efforts. Yet there is an element of doubt and ambiguity that scientists adhere to which in effect conforms to political aims. Future studies could focus on ethical analyses of science from a social perspective whose purpose is to generate insight into the "real" status of science and the validity of the correspondence theory of truth. Nevertheless, science in the hands of those with differing objectives remains a contested boundary.

NOTES

1. Willard Van Orman Quine, *From a Logical Point of View* (Cambridge: Harvard University Press, 1953).
2. The Telegraph (January 27, 2010): http://www.telegraph.co.uk/news/earth/environment/climatechange/7081039/John-Beddington-chief-scientist-says-climate-change-sceptics-should-not-be-dismissed.html.

3. Halil Kilicoglu and Sabine Bergler, "Recognizing Speculative Language in Biomedical Research Articles: A Linguistically Motivated Perspective", *BMC Bioinformatics* 9 (2008): 1–10.
4. Mark Stevenson, Yikun Guo, Robert Gaizauskas, and David Martínez, "Disambiguation of Biomedical Text Using Diverse Sources of Information", *BMC Bioinformatics* 9 (2008): 1–11.
5. Veronica Vincze, György Szarvas, Richárd Farkas, György Mora and János Csirik, "The BioScope Corpus: Biomedical Texts Annotated for Uncertainty, Negation and Their Scopes", *BMC Bioinformatics* 9 (2008): 1–9.
6. Ann Kinzig and David Starrett. "Coping with Uncertainty: A Call for a New Science-Policy Forum", *AMBIO: A Journal of the Human Environment* 32.2 (2003): 330–335 (p. 330).
7. Willard Van Orman Quine, *From a Logical Point of View* (Cambridge: Harvard University Press, 1953).
8. Susan Leigh Star, "Scientific Work and Uncertainty", *Social Studies of Science* 15.3 (1987): 391–427 (p. 392).
9. Karin Knorr-Cetina, in *Epistemic Cultures: How the Sciences Make Knowledge* (Cambridge: Harvard University Press, 1999), described how molecular biology is a science entrenched in social context and symbolic knowledge. Bruno Latour and Steven Woolgar wrote the seminal introduction to social construction of scientific facts in *Laboratory Life* (Princeton: Princeton University Press, 1979). Both of these works are considered pivotal in understanding sociology of scientific knowledge.
10. Kathleen Jordan and Michael Lynch, "The Dissemination, Standardization, and Routinization of a Molecular Biological Technique", *Social Studies of Science* 28.5–6 (1998): 773–800, wrote of the "local management of technique" in their discussion of the molecular biological technique polymerase chain reaction.
11. Bruno Latour and Françoise Bastide, "Essaie De Science Fabrication Mise En Evidence Experimentale Du Processus De Construction De La Realité Par L'application De Methodes Socio-sémiotiques Aux Textes Scientifiques", *Etudes Francaises* 19 (1983): 111–126.
12. Susan Leigh Star, "Scientific Work and Uncertainty", *Social Studies of Science* 15.3 (1987): 391–427 (p. 392).
13. Jakob Jensen, "Scientific Uncertainty in News Coverage of Cancer Research: Effects of Hedging on Scientists' and Journalists' Credibility", *Human Communication Research* 34 (2008): 347–369 (p. 347).
14. Stephen C. Zehr, "Public Representations of Scientific Uncertainty About Global Climate Change", *Public Understanding of Science* 9 (2000): 85–103.
15. *Ibid*, 87.
16. *Ibid*.
17. The New York Times Magazine (3 February 1991): 24–31.
18. *Ibid*, 24.
19. The New York Times (30 January 1995): http://www.nytimes.com/1995/01/30/us/cloud-data-cast-doubt-about-atmosphere.html.
20. The New York Times (17 January 1989): http://www.nytimes.com/1989/01/17/science/global-warming-experts-ponder-bewildering-feedback-effects.html.
21. The New York Times (7 February 1989): http://www.nytimes.com/1989/02/07/science/with-cloudy-crystal-balls-scientists-race-to-assess-global-warming.html.
22. Environment (30 July/August 1988), inside cover.
23. Science 251 (1 March 1991), 1002.
24. William K. Stevens, "At Meeting on Global Warming, U.S. Stands Alone", The New York Times (10 September 1991): C, 1.

25. Stephen C. Zehr, "Public Representations of Scientific Uncertainty about Global Climate Change", *Public Understanding of Science* 9 (2000): 85–103 (p. 87).
26. *Ibid*, 88.
27. Judith Curry, "Reasoning about Climate Uncertainty", *Climate Change* 108 (2011): 723–732.
28. Rachael Jonassen and Roger Pielke Jr., "Improving Conveyance of Uncertainties in the Findings of the IPCC", *Climate Change* 108 (2011): 745–753.
29. M. Granger Morgan, "Certainty, Uncertainty, and Climate Change", *Climate Change* 108 (2011): 707–721.
30. Carolina A. Adler and Gertrude Hirsch Hadorn, "The IPCC and Treatment of Uncertainties: Topics and Sources of Dissensus", *WIREs Climate Change* 5 (2014): 663–676. doi:10.1002/wcc.297.
31. Publications and news networks cited include the Associated Press, Boston Globe, Christian Science Monitor, Cincinnati Enquirer, Cincinnati Post, CNN, Contracosta Times, Detroit News, Kentucky Post, Las Vegas Sun, Los Angeles Times, Minneapolis Star Tribune, New York Times, Pittsburgh Post-Gazette, Reuters, San Francisco Gate, Wall Street Journal, and Washington Post. William P. Cheshire, "Human Embryo Research and the Language of Moral Uncertainty", *American Journal of Bioethics* 4 (2004): 1–5 (p. 2).
32. *Ibid.*
33. *Ibid.*
34. Halil Kilicoglu and Sabine Bergler, "Recognizing Speculative Language in Biomedical Research Articles: A Linguistically Motivated Perspective", *BMC Bioinformatics* 9 (2008): 1–10.
35. These are scientific statements taken from Kilicoglu and Bergler, "Recognizing Speculative Language in Biomedical Research Articles: A Linguistically Motivated Perspective", *BMC Bioinformatics* 9 (Suppl 11) (2008): S10 (p. 2).
36. George Lakoff, "Hedges: A Study in Meaning Criteria and the Logic of Fuzzy Concepts", *Chicago Linguistics Society Papers* 8 (1972): 183–228.
37. Frank Robert Palmer, *Mood and Modality* (Cambridge: Cambridge University Press, 1986).
38. Wallace Chafe, "Evidentiality in English Conversation and Academic Writing", in *Evidentiality: The Linguistic Coding of Epistemology*, eds. Chafe Wallace and Nichols Johanna (Norwood: Ablex Publishing Corporation, 1986), 261–272.
39. Ken Hyland, *Hedging in Scientific Research Articles* (Amsterdam: John Benjamins, 1998).
40. Halil Kilicoglu and Sabine Bergler, "Recognizing Speculative Language in Biomedical Research Articles: A Linguistically Motivated Perspective", *BMC Bioinformatics* 9 (2008): 1–10.
41. Mark Stevenson, Yikun Guo, Robert Gaizauskas and David Martinez, "Disambiguation of Biomedical Text Using Diverse Sources of Information", *BMC Bioinformatics* 9 (2008): 1–11.
42. Veronica Vincze, György Szarvas, Richárd Farkas, György Mora and János Csirik, "The BioScope Corpus: Biomedical Texts Annotated for Uncertainty, Negation and Their Scopes", *BMC Bioinformatics* 9 (2008): 1–9.
43. Alfred Tauber, *Science and the Quest for Meaning* (Waco: Baylor University Press, 2008), p. 96.
44. *Ibid.*
45. Willard Van Orman Quine, *From a Logical Point of View* (Cambridge: Harvard University Press, 1996), p. 96.
46. Alfred Tauber, *Science and the Quest for Meaning* (Waco: Baylor University Press, 2008), p. 97.
47. Joseph Harmon and Alan Gross, *The Craft of Scientific Communication* (Chicago: University of Chicago Press, 2009), p. 50.

Part 2

The Creation of Scientific Terminology

7 Reading Astrolabes in Medieval Hebrew

Josefina Rodríguez Arribas

1. THE TRANSLATION MOVEMENT UP TO THE TWELFTH CENTURY[1]

The twelfth century witnessed an impressive production of scientific treatises in Hebrew, which were the result of the close interaction of Jewish culture with Arabic culture in Muslim territories and the role of Jews in Muslim and Christian courts. Often, Jews were also intermediaries in the intensive transfer of knowledge from Arabic into Latin that most notably took place in the Iberian Peninsula, especially in the twelfth and thirteenth centuries. As a result of this close contact with Arabic culture (and Greek culture, particularly in Sicily),[2] the Hebrew language underwent a process of specialization which saw the creation of many new technical terms. This process first became noticeable in the Iberian Peninsula with Abraham bar Hiyya (Barcelona, d. after 1136) and Abraham ibn Ezra (Tudela, 1089/92–1164/67), and it continued in southern France in the following century. The scientific writings of these authors and translators made medieval Hebrew a language capable of disseminating science and, eventually, a language of scientific research. As this book does not address Medievalists and Hebraists specifically, I would like to make some clarifications that may help readers understand the context and the state of medieval Hebrew when Ibn Ezra first started to explain the use of astrolabes in this language in the first half of the twelfth century.

The background to the first Hebrew treatises on the astrolabe was the important movement of translations that took place in the East and West between the ninth and the twelfth centuries, and particularly those texts translated and commented upon in the Iberian Peninsula, where writers of Arabic, Hebrew, and Latin found themselves living together (often in the same cities) and interested in the same branches of knowledge.[3] It is known that astronomy and astrology were among the first topics to be translated (from Greek, Syriac, and Persian into Arabic, and later from Arabic into Latin and Hebrew). The stellar sciences (especially astrology) were particularly relevant to everyday concerns and politics, which could account for the interest they awoke in the medieval mentality.[4] Because of the large number of texts translated under the patronage of the caliphs in the House of Wisdom in Baghdad,[5] and the translations and original treatises written in Arabic in different parts of the Muslim territories,[6] the best scientific knowledge available in this period was presented in Arabic, and only those who knew Arabic could benefit from it. Jews in the Iberian Peninsula and in other territories under Muslim rule were learned in Arabic and spoke it in their everyday dealings (unlike those living in Christian territories inside and outside Iberia). They could read Arabic manuscripts and benefited from their close relationships with Muslims in the Andalusian courts in which they worked as secretaries, physicians, and translators. However, at the turn of the eleventh and twelfth centuries, a decisive change took place; Abraham bar Hiyya decided to write treatises devoted to astronomy and astrology in Hebrew.[7] Bar Hiyya, who probably mastered the Arabic sciences under the rule of the Banū Hūd dynasty (1039–1118) of Saragossa, worked for the Christian court in Barcelona and perhaps for the court of Alphonse I of Aragon (1104–1134). He decided to write in Hebrew for the Jews of Catalonia and possibly Provence, the cultural area where he lived and worked. Bar Hiyya's decision introduced a radical change in Hebrew literature and science: the Hebrew language had to be enhanced to accommodate the new knowledge introduced in Arabic, for there were no words for all the new scientific concepts and calculations. In addition, the readership of these scientific writings widened (from Arabic to non-Arabic readers) and became more focused (from a multi-religious readership—including Muslims, Christians and Jews—to an exclusively Jewish one).

It is not known for certain whether Hebrew was a spoken language in the Iberian Peninsula. In all likelihood it was not, but it was certainly read and understood.[8] Therefore it was the language that most Jews shared. Literary Hebrew had been greatly enriched in the previous centuries during the golden age of Hebrew poetry (the tenth to twelfth centuries), which drew from two sources in Sepharad:[9] biblical Hebrew (for the terminology and some of the topics) and Arabic models (for the form and content of the poems). Some scientific terms were already present in many of these poems.[10] However, there was no body of terminology capable of expressing most of the sophisticated notions presented in the technical treatises on astronomy, mathematics, and medicine written in Arabic. The Hebrew language had been developed and expanded mainly in the fields of religion and poetry.

2. MEDIEVAL HEBREW IN THE TWELFTH CENTURY

Abraham ibn Ezra was the main translator and disseminator of Arabic science in Hebrew in the twelfth century, the author of a number of treatises on the astrolabe, and the central subject of this chapter. Ibn Ezra left the Iberian Peninsula in 1140 for unknown reasons and never returned. After this we find him writing in Hebrew for the European Jews, whom he met in the different European cities in which he lived.[11] We are given the impression that in their scientific treatises possibly Bar Hiyya, and certainly Abraham ibn Ezra, tried to reach a level of linguistic competence in Hebrew equivalent to the excellence and beauty achieved in Hebrew poetry.[12] This was a concern of great importance since, in addition to providing Jews with a body of scientific knowledge of their own, it would also contribute to providing them with a distinctive culture that would use Hebrew to write and teach any field of knowledge, whether under Muslim or Christian rule. Bar Hiyya and particularly Ibn Ezra enriched Hebrew by providing new terms mainly for astronomy and astrology, but also for mathematics, medicine, and philosophy. The fortunes of this terminology in later Hebrew were mixed; some of the terms are still in use, and others were discarded soon after their coinage.

All Jewish communities shared the knowledge and the language not only of the books that constitute the Hebrew Bible, but also the oral law contained in the Talmud (Palestinian and Babylonian) and the Mishnah, and the Hebrew of other books from the Jewish religious tradition (Midrash, *Sefer yeṣirah*, different *Baraitot*, early religious commentaries written in Hebrew, etc.). The main sources of these texts were also biblical and rabbinical Hebrew and, to a lesser degree, Greek. Therefore, most of the treasury of Hebrew words available in the twelfth century was either biblical or rabbinical/mishnaic. The school of grammarians of the Hebrew language, which flourished in Sepharad in the tenth to twelfth centuries under the influence of the golden age of Arabic grammarians, was much concerned with biblical Hebrew and its characteristics.[13] For Muslims, the Arabic of the Koran was the highest expression of their language and culture. To balance the overwhelming prestige of Arabic language and culture, Jewish grammarians and Hebrew poets (grammarians, poets, and scientists frequently overlapped in the complex medieval culture) took the Hebrew of the Bible, with its grammatical and syntactical rules, as their model to shape the poetic and literary language they put forward.

This parallelism/emulation (and its political and social implications) is also of paramount importance for understanding the context in which Hebrew emerged as a language of science.[14] Hebrew was not only the vehicle of Hebrew science; it also propelled its development in the shadow of the powerful models put forward by Arabic. Until the twelfth century there were Jewish scientists writing science in Arabic and, more frequently, Judaeo-Arabic, that is, Arabic with Hebrew script.[15] However, after Bar Hiyya and Ibn Ezra, most Jews came to write in Hebrew, or at least had the choice of writing in Hebrew as well as in Arabic, Latin, or a vernacular.[16] The treatises of Ibn Ezra, who wrote on most branches of science, made Hebrew a "living language",[17] broadening its traditional literary and religious scope to include other spheres of learning and thought that had been (or were being) developed in Arabic.

In this context of appropriating and disseminating Arabic knowledge, several treatises on astronomical instruments—particularly astrolabes—were written in Arabic and also in Latin, prior to the emergence of treatises on this instrument in Hebrew. Astrolabes had circulated in Europe since at least the tenth century (starting in Catalonia), and the oldest extant Arabic instrument dates back to the beginning of that century.[18] However, the main movement of translations from Arabic into Latin and the composition of original Latin writings about the astrolabe occurred in different parts of Europe a little later, between 1140 and 1150. Among the Latin translators and the original writers of twelfth-century manuscripts on the astrolabe we find Hermann of Carinthia (1143), Raymond of Marseilles (1141), John of Seville (twelfth century), Plato of Tivoli (1133–1145), Robert of Chester (1147 or 1150), an unknown Arialdus (twelfth century), Rudolph of Bruges (1144 or after), and Adelard of Bath (1149–1150).[19] Abraham ibn Ezra's texts on the astrolabe must be placed in the context of this wave of texts and translations on the astrolabe in Christian Europe, which had its roots in the reception and stimulus of scientific texts in Arabic.[20] Abraham ibn Ezra wrote four treatises on the astrolabe: three treatises (*Sefer keli ha-neḥošet*) in Hebrew (Mantua 1146, Verona 1146, and Béziers 1148) and one (*Liber de astrolabio*) in Latin (Rouen 1154).[21] It is possible that the different patrons and reading publics in the four cities in which Ibn Ezra wrote these four treatises (Mantua, Verona, Béziers, and Rouen) demanded slightly different approaches to the same new subject. Ibn Ezra's consideration of the different expectations and interests of the Jews in these cities explains the existence of different Hebrew writings on the same subject within such a short period of time (three treatises between 1146 and 1148). In fact, Ibn Ezra wrote several versions of most of his works.[22] It also may simply have been the case, however, that Ibn Ezra did not have his manuscripts with him when he moved from one city to another and had to write (or dictate) his treatises afresh.

The most important feature of these treatises and their different versions is that they are not in Arabic, the usual language of the Jews in the Iberian Peninsula living under Muslim rule, but in Hebrew (and a few of them also in Latin).[23] Since Jews spoke the language of the peoples among whom they lived, in the Iberian Peninsula they naturally spoke Arabic when they lived under Muslim rule (as they did in Egypt or in the Near East), and later on, when living among Christians (as they did in France), forms of the local vernacular. In both cases, they always kept Hebrew as the language of their tradition and religious customs (which also included religious poetry). However, we must bear in mind that Jews also used Hebrew to communicate with other Jews who spoke a different language (Arabic or a vernacular). Hebrew was the language that most of them knew because all Jews learned to read and write Hebrew in their childhood. The first Jews to write extensively on Arabic science in Hebrew—Bar Hiyya and Ibn Ezra—although educated in Arabic environments, were both writing in territories under Christian rule (Catalonia and several European cities, respectively) in which Jews had to use the local vernacular to deal with their lords and neighbours, with the result that this became their everyday language. The decisive question here is why Bar Hiyya and Ibn Ezra preferred Hebrew to the vernacular. There are two evident answers to this: first, for a Jew, Hebrew was more universal than any vernacular, and second, Hebrew was the language of their religion and their tradition. Hebrew was the distinctive Jewish language—indeed the only one in this period.

3. THE EARLIEST INSTRUMENT-BOOKS IN HEBREW: CONTENTS AND TERMINOLOGY

Ibn Ezra's treatises on the astrolabe (see Table 7.1)—the scientific instrument denoted in Hebrew by the expressions *keli neḥošet* or just *keli*[24] and, exceptionally, *'aṣṭarlub*[25] (and its variants) or *isṭagninut*[26]—are the earliest examples in Hebrew of "instrument-books", that is, texts devoted to the explanation of scientific instruments. Books on astrolabes usually have their contents arranged into three parts: introduction (optional), enumeration and description of the instrument's components, and explanations of the various applications of astrolabes. Table 7.2 shows some of the most important technical terms associated with the components of the astrolabe in Ibn Ezra's three Hebrew treatises.

The section on the applications of the astrolabe is divided into three parts: astronomy, astrology, and land surveying (i.e., how to measure with an astrolabe the distance, height, and depth of buildings, trees, mountains, houses, etc.). The latter is optional, and its position in the treatises fluctuates. Tables 7.3 and 7.4 display some of the technical terminology relevant to the astronomical and to the astrological uses of astrolabes.

The language of these treatises is notable for the use of words and combination of words that already existed to denote the new names related to the instrument. Terms that are familiar from Ibn Ezra's other writings are apparent, for instance: *mešartim* (the planets), *maḥberet* (conjunction), *ha-qav ha-šaveh* (the east-west axis on the plates), *gavhut* (apogee), *neṭiyyah* (solar declination), *tequfat Nisan* (vernal equinox), and *roš ha-Teli* and *zenav ha-Teli* (northern and southern nodes of the planets).[27] However, we also find more unfamiliar terminology in Ibn Ezra's texts: *ta'* (limb), *ma'eṣar / ma'eṣor* and *beriaḥ* (alidade), *meqom ha-roš* (zenith), *ha-qav ha-rišon* (denoting, ambiguously, the first almucantar and also the upper meridian), *šaḥarit* (diurnal twilight) and *'arvit* (nocturnal twilight), *tequfat molad* (the monthly revolution of the moon), *qav ha-šefel* or *ha-qav ha-šafal* (lower meridian), and *meqomot ha-ḥayyim* and *sarim* (aphetic or anairetic places in a horoscope, i.e., positions related to life and death).[28] The terminology is inconsistent in a few cases; notably, the different Hebrew words used to denote the alidade or the sighting device (*zero'a* or *'ezro'a*, *miqzo'a* or *miqṣo'a*, and *ṣela'*) in the third version. There is no way to establish to what extent Ibn Ezra coined this technical terminology himself and to what extent he borrowed terms already in circulation. For instance, in the third version, Ibn Ezra employs the biblical expression *hešev 'afuddah* (the band or girdle of a dress) *in hešev 'afuddat galgal ha-mazzalot* to denote the ecliptic or solar path through the zodiac (literally, "the band or girdle of the dress of the zodiac"), which Solomon ibn Gabirol (Malaga, 1020–158) introduced in his cosmological writing *Keter malkut*.[29]

Other distinctive terms, which, as far as I know, emerge for the first time in Ezra's texts, are: *qodqod* ("crown of the head") for zenith and *qav qodqod* for azimuthal line; *gešer* (bridge) for almucantar or circle/line of altitude; *nihug* (from the Hebrew root 'to direct/guide') for astrological direction; *sevakah* ("network") for the web; and *kavod* (glory/dignity) for the magnitude of a star. For the concept of nadir (the degree diametrically opposing the position of the sun in the sky), he uses *nokaḥ* and *nokaḥ ha-šemeš* (from the Hebrew root meaning to oppose/strike). Ibn Ezra also introduces the distinction between latitude (the distance in relation to the ecliptic) and declination (the distance in relation to the equator). However, they are both denoted inconsistently by *roḥav* and *merḥav* (and declination also by *neṭiyyah*), which implies that the two former terms were not yet completely established.[30] When Ibn Ezra wishes to avoid a potential misunderstanding in his treatises, he does not resort to Arabic or to the vernacular, but to Hebrew (with rare exceptions).[31] The presence of Hebrew synonyms and definitions and the almost nonexistent Arabic, Latin, or vernacular loan words confirm Ibn Ezra's commitment to the Hebrew language.

4. IBN EZRA'S STANCE VIS-À-VIS THE STATUS OF THE HEBREW LANGUAGE

Before considering Ibn Ezra's strategies regarding the coining of technical terms in Hebrew, we should pay some attention to the introductions of his astrolabe treatises, in which he reflects briefly but meaningfully on the Hebrew language. Ibn Ezra opens the second Hebrew version of the treatise on the astrolabe with a short statement on the difficulty of translating into Hebrew:

> Only with great effort do experts have the strength to translate into another language, for we do not know from the holy language but the things written in the Torah. The astrolabe has many names but I will mention them according to what is within the reach of my knowledge.[32]

In this fragment Ibn Ezra states two things: first, that the Jews of his time had only the remnants of an old language (biblical Hebrew) which had been richer in a distant past; second, that there were many words that co-existed in his time to describe astrolabes (that is, many traditions merging in medieval Hebrew to denote the new knowledge taken from Arabic). In the third version of his treatise, he makes his most eloquent statement regarding the Hebrew language:

> Anyone paying attention knows that the holy language was broader than all the gentile languages, for it was the oldest language. Because

> of the exile of the holy people from the holy land, they mixed with the gentiles, learnt their languages, and forgot to speak their own language. They were left only with the stories of the prophets and we do not find anything written about those things they did not need to talk about. It is difficult to create words and that is why people transliterate words from one language to another. This treatise is called *'isṭagninut*,[33] but I called it *keli neḥošet* [i.e., the instrument of brass], for most of them are made of brass.[34]

In Ibn Ezra's view, Jews, in their forced diaspora, had forgotten their own language and almost all the knowledge that had once been diffused in Hebrew; only the words of the holy writings remained. However, many biblical words offered powerful clues about the scope of Hebrew and its past semantic richness. In order to understand the status of Hebrew in Ibn Ezra's view, let us consider briefly a key biblical term. In his grammatical treatise *Sefer ṣaḥot* (*The Book of the Purity* [of Language]), Ibn Ezra classifies the letters of the Tetragrammaton, the most mysterious name of God (Y-H-W-H), as palatal (*yud*), guttural (*he*), and interdental (*waw*).[35] According to Ibn Ezra, these letters are (with *'alef*) the principal letters of the Hebrew alphabet, since they are the letters of all the names of God that are nouns (*Yah*, *'Ehiyeh*, *Yhwh*) as opposed to those names that are adjectives (e.g., '*El-Šaddai* or '*Elohim*, among others).[36] Ibn Ezra applies to these letters analogical thought, one of the main features of the medieval mind-set. The law of analogy worked as a sort of matrix by which to move horizontally and vertically through the medieval universe and the beings that populated it. This matrix was able to interconnect everything in the universe and to generate new meanings from these connections. Ibn Ezra's application of analogy to the letters *yud*, *he*, and *waw* shows us that the shapes of these letters are related to their meanings, sounds, and numerical values and points to a double reality—one that is both physical and theological.

Concerning the shapes of these letters, Ibn Ezra explains the correspondence of *yud* with the uppermost sphere (the tenth in his astronomical system), which encircles the two main motions in the heavens (the westward or diurnal motion of the stars and the eastward motion of the planets).[37] The letter *he* seems to refer to that cosmological instant when matter receives form and individuals emerge. It is known that Ibn Ezra, in his conception of the universe, supported hylomorphism (the intrinsic union of a substantial form and a primary matter) for all the beings up to the eighth sphere; only the two upper spheres (ninth and tenth), which constitute the upper world, were bodiless (in all likelihood, immaterial).[38] In his explanation of this letter, Ibn Ezra seems to suggest that every form has a suitable matter and *vice versa*, each in the way God established it at the Creation. The third letter of the Tetragrammaton—*waw*—represents the axis of the sphere whose axis of rotation is shared by all the spheres in the universe (the diurnal motion).[39] In conclusion, the Tetragrammaton reveals the divine presence in three elements of the universe: first in the encircling sphere of the world; second in the link between form and matter; and third in the unique motion shared by all the spheres. It seems that Ibn Ezra believed that human language was of divine origin (although not revealed by God) and that Hebrew was the primeval language, from which the other Semitic languages (such as Aramaic and Arabic) derived. According to Ibn Ezra, the Hebrew language was a reality that had been somehow enmeshed in the order of the world since its very beginning, and Adam had been intelligent enough to find the right words with which to name animals, plants, and the other things he encountered in the Garden of Eden (for there were "right names" for everything created).[40] Are the names of God the only case of adequacy in Hebrew between beings and their names? How is the relationship between words and meanings built where Ibn Ezra's technical terms for astrolabes are concerned? Let us return to Ibn Ezra's technical terminology in his Hebrew treatises on the astrolabe to find some clues that may help us answer these questions.

5. IBN EZRA'S STANCE VIS-À-VIS THE COINAGE OF TECHNICAL TERMS IN MEDIEVAL HEBREW

Perusal of the technical terms emerging in the three Hebrew treatises that Ibn Ezra devoted to the explanation and uses of the astrolabe reveals that he employed two primary strategies in coining terminology.[41] The first strategy involved systematic calques (loan translations) of Arabic terms, for which Ibn Ezra provides the equivalent Hebrew term (see Table 7.5). The second strategy involved either the recovery of existing Hebrew terms that had a distant or close semantic relationship to the meanings they denote or the creation of word pairs semantically related to the meanings intended (see Table 7.6). The second strategy (with its two sub-strategies) is especially interesting, for it provides an opportunity to appreciate how Ibn Ezra played with words (and their biblical resonances) and meanings when he combined his knowledge of biblical words with his proficiency in Arabic astronomy and astrology. This interplay affirms the high esteem in which he held biblical Hebrew. If we consider the terms in both tables (7.5 and 7.6), the proportion of biblical terms and biblical roots is overwhelming, whether used alone or combined in non-biblical expressions, such as *ta'*, *ḥeq*, *qodqod*, *sevaḵah*, *yated*, *rešet*, *goral*, *gešer ri'šon*, and *Tannin mešaret*. The few non-biblical words in Tables 7.5 and 7.6 are *saddan*, *neṭiyyah*, *tiqqun*, and *Sarṭan*. The Arabic loan word *qoṭev* (Table 7.6) had entered Hebrew before the time of Ibn Ezra, who used it as a synonym for the Hebrew word *saddan*, as Abraham bar Hiyya had done.[42] It must also be mentioned that the confusion between *yated* (angle) and *yeter* (chord) already existed in Bar Hiyya's astronomical texts and continued occasionally in Ibn Ezra's manuscripts.[43] In all likelihood, this confusion must have taken place among the medieval copyists.[44] A review of these terms in the treatises on the astrolabe suggests that Ibn Ezra found many technical terms already in use when he started to write on the instrument, notably calques of Arabic words and expressions such as the previously mentioned *qoṭev* or *sevaḵah*. He could not avoid an Arabic loan word like *qoṭev*, which Bar Hiyya, his predecessor, had used. However, when he felt free to do so, he chose biblical terms and original Hebrew words to translate the Arabic knowledge on the astrolabe.

As mentioned earlier, biblical language was to medieval Hebrew what Arabic sciences were to medieval knowledge: a model to follow and to assimilate. In this way, resorting to biblical terminology to coin new words would not be unexpected among some medieval writers and translators. What makes the poet and grammarian Ibn Ezra a purist of the Hebrew language, and thus distinct from Bar Hiyya, is his frequent (and deliberate) use of biblical terms in the technical vocabulary of his scientific treatises, of which the treatises on the astrolabe form a representative sample. One likely reason for the predominance of biblical words in Ibn Ezra's technical vocabulary is that he wrote some of his biblical commentaries—which contained numerous excursuses on scientific matters, particularly on astrology and astronomy—before he started to develop his scientific writings.[45] Therefore he had already coined some of the technical terms that he would subsequently employ in his monographs on the science of the stars. Hence a preliminary conclusion one could draw would be that technical terms had already emerged as an important part of the terminology he used in his early biblical commentaries and were then properly established as such in his scientific writings (a reason having to do with the chronology of his texts).[46]

Another possible reason is the Arabic model, in which the language of the Koran was the model (a reason having to do with the social and political predominance of Arabic culture). There is still a third reason, one of a religious and political nature: with his terminological choices, Ibn Ezra was voicing an old belief among the Jews of Sepharad. If he found terms for these sciences in the Torah, it is because he believed that the patriarchs and prophets had known them and that they had been part of the civilization of Israel since its origin, as others before him, such as Yehudah ha-Levi (d. 1141) or Abraham bar Hiyya, also believed.[47] If Ibn Ezra had to restore to these words their original scientific meanings, it was because his contemporaries, and before them their ancestors, had forgotten them. Consequently, he did not consider the work of re-allocating meaning and scientific import to some words of the Torah an arbitrary task. Apparently Ibn Ezra did not believe he was creating new vocabulary from ancient biblical words; instead he was convinced his task was to recover and restore the original meanings of the biblical words. He alludes implicitly to the idea that the holy language has lost something through time; the biblical words have lost their meanings, which he is willing to recover by means of the right exegesis. Ibn Ezra was after all one of the most important medieval interpreters of the Bible, and his exegesis of the biblical text is not only a restoration of meanings to the Torah, but also a restoration of the Hebrew language (syntax, grammar, and vocabulary), the task of a grammarian he also was.[48] In addition, with his vindication of Hebrew as a language of science, Ibn Ezra helped reinforce a central cultural movement of his time, a period of renaissance for the Hebrew language, when Jews "made Biblical Hebrew the Jewish equivalent of classical Arabic and accorded their language a new status as a cultural monument above and beyond its traditional status as the holy tongue".[49]

The way in which Ibn Ezra followed Arabic patterns in the sample of terms presented in Table 7.5 has much to do with the way Arabic translators had perceived certain patterns in the Greek, Persian, and Syriac technical words they translated into Arabic.[50] We will consider here just a few Hebrew terms and their immediate Arabic models (a more in-depth approach is outside the scope of this study). *Almucantar* in both languages has a root that conveys the image of something joining or making contact between two sides or distant points—such as an aqueduct or a bridge. This image makes sense since an almucantar is a curve joining the two opposite sides of the upper half of any northern astrolabe.[51] The concept of *sphaera recta* is also, in both languages, a translation of the visual representation of this term: an observer at the equator, facing east, would see the celestial equator rising perpendicularly in relation to the horizon. In this way, *mustaqīm* and *yošer* mean "what makes straight, upright". The procedure by which to calculate astrological houses is denoted in Hebrew by an expression similar (although not identical) to the Arabic one. The Arabic term *taswiyat* denotes a procedure that seems to imply some sort of equivalence between astrological houses, while the Hebrew *tiqqun* indicates some kind of rectification or correction made in order to determine the houses. In any event, they both relate to the fact that, with the exception of the four angles (the rising sign, the setting sign, the upper mid-heaven, and the lower mid-heaven), whatever the sizes of the remaining eight houses, any two of them opposite each other in the zodiac will be equal (*taswiyat al-buyūt*).

In Table 7.6, I have included Hebrew terms that represent a "Hebrew shift" with respect to the Arabic models, as the meanings given in parentheses demonstrate: *Tannin* is a biblical expression that was made the equivalent of the Persian *ǧawzahar*, the pseudo-planet—represented as a dragon—which was responsible for the disappearance of the moon every month according to Indian and other ancient traditions. The same meaning is conveyed by the post-biblical word *Teli*, which emerged for the first time in Hebrew in the *Sefer yeṣirah*.[52] *Sahm* and *goral* both refer to fortune as the result of casting lots (*sahm*). The Arabic term means arrow—referring specifically to the old custom of foretelling the future with arrows, according to the way they moved through the air and fell on the ground.

6. CONCLUSIONS

Iberian Jews were interested in astrolabes; they used them and wrote about them. As Arabic became progressively more of a foreign language for them because of the diaspora moving from the Muslim territories, more translations were required. In addition, many realised that knowledge of Arabic science was not available to most Jews living outside Iberia. Consequently, Jews from the Iberian Peninsula decided to make the instrument and the texts concerning it available in their own language, for any Jew anywhere could read and understand Hebrew. Therefore Hebrew was used to write texts on the astrolabe, and it also, occasionally, became the script used on some scientific instruments—as the astrolabe with Hebrew and Judaeo-Arabic inscriptions in the medieval collections of the British Museum.[53] These circumstances led, in the twelfth century, to the flourishing of scientific literature in Hebrew, written by Jews who lived or had lived in al-Andalus, for they were familiar with the Arabic language and the Arabic sciences. Abraham ibn Ezra was the first to introduce the astrolabe in Hebrew. I have analysed here some of his terminological choices in coining new terms or in giving new meanings to pre-existing words in order to denote and describe the astrolabe, its components, and its applications. In his treatises, Ibn Ezra introduced many words with biblical roots for three possible reasons. First, the prestige of the Arabic language among Jews and the status of the Koran among Arabic writers, which encouraged Ibn Ezra and others to aspire to an equivalent level in Hebrew in all fields of knowledge. Second, the fact that Ibn Ezra had previously explained scientific matters in his early biblical commentaries before writing his texts devoted exclusively to sciences. This is relevant because these scientific explanations occurred in biblical contexts and necessarily implied the use of biblical words in the exegesis of the biblical text. Finally, the medieval Jewish myth of a science in Hebrew that pre-dated Greek and Arabic sciences. However, many questions remain unanswered about Ibn Ezra's stance vis-à-vis the Hebrew language, especially because of his elusive and concise style and his numerous writings.

Table 7.1 Ibn Ezra treatises on the astrolabe

Treatises	Dates	Places
Hebrew 1st version	1146	Mantua (Italy)
Hebrew 2nd version	1146	Verona (Italy)
Hebrew 3rd version	1148	Béziers (France)
Latin	1154	Rouen (Normandy)

Table 7.2 Some technical terms for the components of the astrolabe[54]

Term	Heb. 1[1]	Heb. 2[2]	Heb. 3[3]
Azimuthal arc	*qav qodqod* / *'al-samt*[4]	*qav qodqod*	*gešer* / *gešer qav qodqod*
Pointer of a star	*šen ha-kokav*	*šen ḥad* / *šen ha-kokav* / *ḥod* / *šen*	*ḥod* / *ḥaddud*
Almucantar or arc/ circle of latitude	*gešer*	*gešer*	————
Arc (of a circle)	*qešet*	*qešet*	*qešet*
Tropic of Cancer	*ro'š Sarṭan* / *sof ṣafon*	*'agullat ro'š Sarṭan* / *ro'š Sarṭan* / *sof ṣafon*	*'agullat ro'š Sarṭan* / *sof ṣafon* / *ro'š Sarṭan*
Tropic of Capricorn	*ro'š Gedi* / *sof drom*	*'agullat ro'š Gedi* / *ro'š Gedi* / *sof drom*	*'agullat ro'š Gedi* / *sof drom* / *ro'š Gedi*
Indicator (*almuri*)	*ro'š Gedi*	*ro'š Gedi*	*šen*
Digit	*'eṣba'*	*'eṣba'*	*'eṣba'*
Alidade	*ma'eṣar* or *ma'eṣor* / *beriaḥ*	*ma'eṣar* or *ma'eṣor*	*miqṣo'a* / *ṣela'* / *'ezro'a*
Seasonal or uneven hour	*ša'ah me'uvvet*	*ša'ah me'uvvet*	*ša'ah me'uvvet*
Upper meridian	*qav ḥeṣi ha-šamayim* / *qav ḥeṣi ha-yom*	*qav ḥeṣi ha-šamayim* / *ḥeṣi ha-šamayim*	*qav ḥeṣi ha-šamayim* / *qav ha-rom*
Lower meridian	*qav yetad ha-'areṣ* / *qav yeter ha-'areṣ* / *'emṣa' ha-'areṣ* / *qav ha-tehom*	*qav yetad ha-'areṣ* / *qav ha-šefel*	*qav ha-semo'l* / *qav taḥat ha-areṣ* / *qav taḥat ha-tehom* / *qav ha-tehom*
Limb of the mother	*ta'*	*ta'*	*ḥeq*
Mother	*'em*	*'em*	*'em*
Sighting vane	*luaḥ ribua'*	*luaḥ*	*luaḥ*
Shadow square	*merubba' ha-ṣelalim*	*ma'alah*	*ṣurah* / *merubba' ha-ṣel*
Spider or rete	*rešet* / *sevakah*	*rešet* / *'agullat ha-mazzalot*	*rešet* / *sevakah*
Plate	*luaḥ* *bat*	*luaḥ*	*luaḥ*
Oblique shadow	*ṣel hafuk*	*ṣel hafuk*	*ṣel hafuk*
Right shadow	*ṣel yašar*	*ṣel yašar*	*ṣel yašar*
Zodiac	*galgal ha-mazzalot*	*galgal ha-mazzalot*	*galgal ha-mazzalot*

(*Continued*)

Table 7.2 (Continued)

Term	Heb. 1[1]	Heb. 2[2]	Heb. 3[3]
Ecliptic or solar path	*'agullat ha-mazzalot / 'afuddat ha-galgal / qav 'afuddat ha-galgal / qav 'afuddat galgal ha-mazzalot*	*'agullat ha-mazzalot / 'afuddah / qav galgal ha-mazzalot / 'afuddat ha-galgal / 'afuddat galgal ha-mazzalot*	*ḥešev 'afuddat galgal ha-mazzalot / ḥešev 'afuddah*
Wedge	*sus*	————	*sus*
Equator	*'agullat mašveh ha-yom*	*qav yošer*	————
First almucantar	*gešer ri'šon / 'ofeq*	*gešer ri'šon / qav ri'šon*	*gešer ri'šon*
Division of hour		*qav ha-ša'ah*	*qav ha-ša'ah / qav tehillat ha-ša'ah*
Azimuth	*samt*	*qodqod*	*qodqod*
Zenith	*'emṣa' ha-šamayim*	*ro'š*	————
Lunar mansion	*maḥaneh ha-levanah*	*maḥaneh ha-levanah*	————

[1] MS Bibliothèque Nationale de France 1061, ff. 148a-163b, MS Bibliothèque Nationale de France 1053, ff. 1a-36a, and H. Edelman (faulty Hebrew ed.) (Könisberg: Meir Ben Yithaq Bakal 1845).
[2] MS Bibliothèque Nationale de France 1045, ff. 187b-196a.
[3] MS Moscow, Gunzburg 937, ff. 2a-13b.
[4] *Al-samt* is an Arabic loan.

Table 7.3 Some technical terms for the astronomical applications of astrolabes

Term	Heb. 1	Heb. 2	Heb. 3
Rise/ascension	*'alut / ziruaḥ / 'aliyyah*	*'alut / ziruaḥ*	*'alut*
Orb/sphere/circumference	*galgal*	*galgal*	*galgal*
Right sphere	————	*galgal ha-yošer*	*galgal ha-yošer / ha-galgal ha-'elion*
Co-equalize		*le-taqqen*	*le-taqqen*
Chord (trigonometry)	*yeter*	*yated / yetad ha-qešet*[1]	————
Equal degrees/degrees of sphaera recta	————	*ma'alah yešarah*	*ma'alah yešarah*

(*Continued*)

Table 7.3 (Continued)

Term	Heb. 1	Heb. 2	Heb. 3
Latitude/declination	*roḥav*	*roḥav / merḥav*	*roḥav / merḥav*
Noon	*ḥeṣi ha-yom*	*ḥeṣi ha-yom*	*ḥeṣi ha-yom*
Pole	*saddan*	*qoṭev / saddan*	*saddan*
Circle/cycle	*'agullah*	*'iggul / 'agullah / galgal*	*'iggul / 'agullah*
Declination	*neṭiyyah*	*neṭiyyat ha-šemeš*	*neṭiyyah*
Clime	*'ofeq / 'aqlim*	*gevulim ha-arṣot*	*gevul*
Magnitude of a star	*kavod*	*kavod*	*kavod*
To move any planet or the sun eastward or westward To incline the sun northward or southward with respect to the equator	*naṭah*	*naṭah*	*naṭah*
Fixed star / Superior planet	————	*koḵav 'elion*	*koḵav 'elion*
Inferior planet	————	*koḵav šafal*	*koḵav šafal*
Longitude	————	*'oreḵ*	*'oreḵ*
Altitude of the pole	————	*govah ha-qoṭev*	————
Solar declination	*neṭut ha-šemeš*	*neṭut ha-šemeš*	————
Minimal altitude of a star	*šiflut*	*šiflut*	————
Zodiacal latitude of a fixed star	*merḥav*	*merḥav*	————
Northern node of a planet	————	*Tannin mešaret*	————
Southern node of a planet	————	*zenav mešaret*	————
Arc of visibility	*qešet ha-mar'ah*	*qešet ha-mar'ah*	————
Mean motion of a planet or luminary	————	*mahalaḵ 'emṣa'i*	————
Anomaly of a planet or luminary	————	*manah*	————
Deferent	————	*galgal ha-muṣaq*	————
Apogee	————	*gavhut*	————
Perigee	————	*šiflut*	————
Epicycle	————	*galgal qaṭan*	————
Northern node of the moon	————	*ro'š ha-teli*	————
Southern node of the moon	————	*zenav ha-Teli*	————
The diametrically opposite degree or nadir	*nokaḥ / nokaḥi*	*nokaḥ*	*nokaḥ*

[1] Common confusion of *yated* and *yeter*.

Table 7.4 Some technical terms for the astrological applications of astrolabes

Term	Heb. 1	Heb. 2	Heb. 3
To direct	———	*li-nhog / le-naheg*	*li-nhog*
Astrological direction	———	*nihug*	*nihug*
Rising-sign	———	———	*mazzal ṣomeaḥ / ṣomeḥet*
Astrological aspect	*mabbaṭ*	*mabbaṭ*	*mabbaṭ*
Rising degree (3rd and 2nd v.) / First astrological house (2nd v.)	*ma'alah ṣomeḥet*	*ma'alah ṣomeḥet*	*ma'alah ṣomeḥet / ma'alah zoreḥet*
The descending degree or descendant	*ma'alah šoqa'at*	*ma'alah šoqa'at*	*ma'alah šoqa'at*
The first astrological house	*ha-yated ha-mizraḥi*	———	*ha-yated ha-mizraḥi*
The tenth astrological house	*yated 'elion*	*yated 'elion*	*yetad ha-drom / yated dromi / yetad ha-rom*
The fourth astrological house	*yetad ha-taḥton*	*yetad ha-'areṣ / yated revi'i*	*yated šafal / yated semo'li*
The seventh astrological house	———	*yated šivi'i / yated ma'aravi*	*yated ma'aravi*
Annual revolution of the sun / Profectio	———	*tequfah / tequfat ha-šanah / tequfat šenat ha-'olam*	*tequfah*
Northern aspect	*mabbaṭ semo'li*	*mabbaṭ semo'li*	*mabbaṭ semo'li*
Southern aspect	*mabbaṭ yemani*	*mabbaṭ yemani*	*mabbaṭ yemani*
The beginning of month (the revolution of the moon)	———	*tequfat molad*	———
The revolution of a natal horoscope	———	*tequfat šenat molad*	———
Domification[1]	*tiqqun ha-battim*	*tiqqun ha-battim*	———
The division of the ecliptic into the four angles	———	*tiqqun ha-yetedot*	———
Aphetic/anairetic places in a horoscope	———	*meqomot ha-ḥayyim / sarim*	———
Lot of fortune	———	*goral ha-mazzal ha-ṭov*	———

[1] Division of the ecliptic into 12 astrological houses according to date and place.

Table 7.5 Some technical terms following the Arabic pattern

Meaning	Arabic Pattern[1]
Almucantar	*gešer* (bridge) related to the Arabic *muqanṭara* (bridge)
Arc of circle	*qešet* (arc) related to the Arabic *qaws* (arc)
Tropic of Cancer	*ro'š Sarṭan* (the beginning of Cancer) related to *[madār] ra's as-Saraṭān* (cycle of the beginning of Cancer)
Oblique shadow	*ṣel hafuḵ* (oblique shadow) related to *ẓill mankūs* (re-versed shadow)
Zenith	*ro'š* (head) related to *[samt] ar-ra's* (direction of the head)
Spider or rete	*sevaḵah / rešet* (net) related to *šabaka* (net)
Orb/sphere/circumference	*galgal* (orb) related to *kura* or *falaq* (sphere, circle)
Right sphere	*galgal ha-yošer* (right sphere) related to *falaq al-mustaqīm* (right sphere)
Declination/inclination	*neṭiyyah* (declination) related to *maīl* (inclination)
Astrological house	*bayit* (house) related to *bayt* (house)
Rising-sign	*mazzal ṣomeaḥ / ṣomeḥet* (rising sign or ascendant) related to *maṭālīʿ/ ṭālīʿ* (rising point)
Angle of the lower meridian	*yetad ha-'areṣ* (angle of the earth) related to *watad al-'ard* (angle of the earth)
Annual revolution of the sun	*tequfah* (revolution) related to *taḥwīl*[2] (transfer, shift)
Domification	*tiqqun ha-battim* (correction, rectification of the houses) related to *taswiyat*[3] *al-buyūt* (equalization of the houses)
Equator	*qav yošer* (line of rightness) related to *dā'irat al-iʿtidāl* (circle of balance)
Noon	*ḥeṣi yom* (noon) related to *zawāl* (mid-day)

[1] For the Arabic terms see Willy Hartner, "The Principle and Use of the Astrolabe," in Willy Hartner (ed.) *Oriens-occidens* (Hildesheim: Olms 1968), pp. 287–318; Richard Lemay (Arabic and Latin ed.), *Abū Ma'shar's Kitāb al-madkhal al-kabīr ilá 'ilm aḥkām al-nujūm = Liber introductorii maioris ad scientiam judiciorum astrorum*, 9 v. (Naples: Istituto Universitario Orientale 1995); and Paul Kunitzsch, "Glossar der arabischen Fachausdrücke in der mittelalterlichen europäischen Astrolabliteratur," *Nachrichten der Akademie der Wissenschaften in Göttingen, Philologisch-Historische Klasse*, 1982, pp. 459–571.

[2] Meaning: to put something in a place different from the place it was in.

[3] Meaning: to make equal or parallel.

Table 7.6 Technical terms following a Hebrew pattern

Meaning	Hebrew Pattern	Arabic[1]
Azimuth	*qodqod* (head, apex, top of the head)	*samt* [*ar-ras*] (direction [of the head])[2]
Limb or rim	*ta'* (chamber)	*ḥuğra* (side) / *ṭawq* (ring) / *kiffa* (curve)
First almucantar	*gešer ri'šon* (first bridge or first almucantar)	*ufq* / *ufq* al-mašriq *wa'l-maghrib* (horizon/east-west horizon)
Pole	*saddan*[3] (pole)	*qutb* (axle, spindle, axis)
Epicycle	*galgal qaṭan* (the small circle)	*falak al-tadwīr* (sphere causing to turn in a circle)
Vernal and autumnal equinoxes	*tqufat Nisan and tqufat Tišri* (vernal and autumnal equinoxes)	*i'tidāl arrabī' and i'tidāl ḵarīfī* (the balance of spring and the balance of crop/ harvest)
Northern/southern node of a planet	*Tannin ha-mešaret* (the Dragon of the planet)	*ğawzahar al-kawkab* (the Dragon of the planet)
Northern/southern node of the moon	*ro'š and zenav ha-Teli* (the head and the tail of Teli)	*rā's and dhanab ğawzahar al-qamar* (the head and the tail of the Dragon of the moon)
Astrological direction	*nihug* (direction)	*tasyīr* (motion, sending out, projection)
Aphetic place in a horoscope	*meqomot ha-ḥayyim* (place of life) / *sar* (prince)	qāṭi'?
Lot of fortune	*goral ha-mazzal ha-ṭov* (lot of the good fortune)	*sahm al-sa'āda* (arrow of happiness)

[1] For the Arabic terms, see Hartner, "The Principle and Use of the Astrolabe," pp. 287–318; Lemay, *Abū Ma'shar's Kitāb al-madkhal al-kabīr ilá 'ilm aḥkām al-nujūm*; and Kunitzsch, "Glossar der arabischen Fachausdrücke in der mittelalterlichen europäischen Astrolabliteratur." We find also the Arabic loan *al-samt* in the first version.

[2] I.e., the distance of a star measured along the horizon, from east or west.

[3] Also *qoṭev*, as mentioned in the article.

NOTES

1. I would like to express my appreciation to the Frankel Institute for Advanced Judaic Studies at the University of Michigan (Ann Arbor) for providing the collegial context in which a large part of this article was written. I am indebted to the late Ángel Sáenz-Badillos and to Amos Geula for their helful comments.
2. Regarding Hebrew literature and translations in southern Italy and Sicily, see Cecil Roth, "Jewish Intellectual Life in Medieval Sicily," *The Jewish Quarterly Review* 47.4 (1957): 317–335; Elinor Lieber, "Asaf's Book of Medicines: A Hebrew Encyclopedia of Greek and Jewish Magic, Possibly Compiled in Byzantium on an Indian Model," *Dumbarton Oaks Papers* 38 (1984): 233–249, and "An Ongoing Mystery: The So-Called Book of Medicines Attributed to Assaf the Sage," *Bulletin of Judeo-Greek Studies* 6 (1991): 18–25; and Andrew Sharf, *The Universe of Shabbetai Donnolo* (New York: Ktav, 1976). Alex Metcalfe states that Hebrew as the target language of the translations emerged in Italy only in the thirteenth century; see *The Muslims of Medieval Italy* (Edinburgh: Edinburgh University Press, 2009), 259–260.
3. For a complete overview of the translation movement in the Iberian Peninsula (as well as in Italy and Southern France) between the tenth and thirteenth centuries, see José M. Millás-Vallicrosa's article, "Translations of Oriental Scientific Works (To the End of the Thirteenth Century)," in *The Evolution of Science, Readings from the History of Mankind,* eds. Guy S. Métraux and Fraçois Crouzet (Chicago and New York: New American Library, 1963), 128–167.
4. For a bibliography of Bar Hiyya's and A. ibn Ezra's scientific terminology and language, see Josefina Rodríguez-Arribas, "Terminology for Historical Astrology According to A. Bar Hiyya and A. Ibn Ezra," *Aleph* 11.1 (2011): 13, n3 and 41, n55 respectively.
5. The Abbasid dynasty established the House of Wisdom in the city of Baghdad, the new capital, in 762. The House of Wisdom was a centre for the study and translation of the knowledge available at that time in different languages (mainly Persian, Indian, and Greek) and endured until the thirteenth century, when a Mongol invasion destroyed Baghdad (1258). Famous scholars and translators associated with this institution were Hunayn ibn Ishaq (809–873) and Thabit ibn Qurra (826–901), among many others. Regarding the House of Wisdom and the transfer of knowledge around it see, for instance, Dimitri Gutas, *Greek Thought, Arabic Culture: the Graeco-Arabic Translation Movement in Baghdad and Early 'Abbāsid Society (2nd-4th/8th-10th centuries)* (London and New York: Routledge, 1998).
6. Muslim territory in this period extended from India and Persia to Northern Africa and the Iberian Peninsula and included Syria, Palestine, Armenia, the Arabian Peninsula, and Egypt.
7. Some scientific material existed in Hebrew prior to Bar Hiyya's and Ibn Ezra's work, particularly in the fields of medicine and cosmology. See, for instance, Shabbetai Donnolo's *Commentary on the Sefer yeṣirah*, Piergabriele Mancuso [Hebrew ed. and English tr.], *Shabbatai Donnolo's Sefer Ḥakhmoni* (Leiden: Brill, 2010). However, the sources of these works were mainly Greek, not Arabic, and the technical terminology involved was still more cosmological than astronomical.
8. Jews in al-Andalus (i.e., the part of the Iberian Peninsula under Muslim rule) lived in a society where the language of culture and power was Arabic (even though many of the conquerors were North African Berbers speaking Berber languages). Therefore Arabic became the language of most of its subjects

(Christians and Jews), although they also preserved and used their original languages (Latin and Hebrew) in their documents and in religious activities. It is worth remembering that, in the tenth century, when the Caliphate of Cordova reached its apogee, al-Andalus occupied more than two-thirds of the Iberian Peninsula, and in the first period of the conquest, Muslim troops reached Poitiers in the heart of France (732). Regarding the languages spoken by the Jews living in the Iberian Peninsula, see David Romano, "Hispanojudíos traductores del árabe," *Boletín de la Real Academia de Buenas Letras de Barcelona* 43 (1991–1992): 215–216. For a different view, see William Chomsky, "The Growth of Hebrew during the Middle Ages," *The Jewish Quarterly Review New Series* 57 (1967): 121–136; Chomsky holds that "Hebrew has never really ceased to be employed for conversational purposes" (*ibid.* p. 121).

9. Sepharad was the term Jews used to refer to the territories of the Iberian Peninsula where they lived.
10. For astronomical and astrological content in the poetry of Sepharad, see Josefina Rodríguez Arribas, "Science in Poetic Contexts: Astronomy and Astrology in the Hebrew Poetry of Sepharad," *Miscelánea de estudios árabes y hebraicos, sección hebreo* 59.1 (2010): 167–202.
11. Many articles and some books have been devoted to the role of Ibn Ezra as a disseminator of Arabic science in Europe and his contribution to the Renaissance of the twelfth century. For a list of Ibn Ezra's works and bibliography on his work, see Josefina Rodríguez Arribas, "Ibn Ezra, Abraham (Abu Iṣḥāq)," *Encyclopedia of Jews in the Islamic World* (Leiden: Brill, 2012), http://referenceworks.brillonline.com/entries/encyclopedia-of-jews-in-the-islamic-world/ibn-ezra-abraham-abu-is-h-a-q-COM_0010510 (accesed March 7, 2015). Particularly relevant to this study is Ángel Sáenz-Badillos, "Abraham ibn Ezra and the Twelfth-Century European Renaissance," in *Studies in Hebrew Literature and Jewish Culture Presented to Albert van der Heide on the Occasion of his Sixty-Fifth Birthday*, ed. Martin Baasten and Reinier Munk (Dordrecht: Springer, 2007), 1–20. In this article, Sáenz-Badillos considers Ibn Ezra within the more general framework of the European intellectual life of the twelfth century and presents him as a European scholar. As far as we know Ibn Ezra only wrote poetry before leaving Sepharad.
12. For an overview of the poetry of the golden age in medieval Hebrew (tenth through thirteenth centuries), see Peter Cole, *The Dream of the Poem: Hebrew Poetry from Muslim and Christian Spain, 950–1492* (Princeton and Oxford: Princeton University Press, 2007).
13. Regarding Hebrew grammars and Jewish grammarians, see Ángel Sáenz-Badillos and Judit Targarona Borrás, *Gramáticos hebreos de al-Andalus* (Córdoba: El Almendro, 1988). Regarding Hebrew language, see Sáenz-Badillos, *A History of the Hebrew Language* (Cambridge: Cambridge University Press, 1993), especially 202–264 (medieval Hebrew).
14. Jews were in close contact with the impressive Arabic culture displayed in Muslim courts: poetry, music, philosophy, language, science; all fields were affected by the Golden Age that Muslims enjoyed. Even when Jews were living among Christians and under Christian rule, Arabic learning was also a model, for many Christians in Europe were also undertaking the task of translating and assimilating Arabic knowledge. At the same time, Arabic was the expression of a people who practised a religion different from, and in many aspects contrary to, the Jewish one, but above all, Muslims were the rulers, those holding power. As Jews did not have a scientific or artistic culture of their own, they appropriated the culture of their rulers while at the same time trying to keep their own tradition and identity as a people. Bar Hiyya's and Ibn Ezra's introduction of Arabic sciences into the Hebrew language was a

decisive movement in this process of acculturation. In relation to the Jews speaking Arabic and the challenge it implied, see Robert Chazan, *The Jews of Medieval Western Christendom, 1000–1500* (Cambridge: Cambridge University Press, 2006), 25. See also Abraham S. Halkin, "The Medieval Jewish Attitude toward Hebrew," in *Biblical and Other Studies,* ed. Alexander Altman (Cambridge, Massachusetts: Harvard University Press, 1963), 233–248; Nehemiah Allony, "The Reflection of the Revolt against 'Arabiyyah' in Medieval Jewish Literature," in *Sefer Meir Wallenstein on the Occasion of His Seventy-Fifth Birthday*, eds. Chaim Rabin et al. (Jerusalem: Kiryat Sepher, 1979), 80–136 [Heb.]; Norman Roth, "Jewish Reactions to the ʿArabiyya and the Renaissance of Hebrew in Spain," *Journal of Semitic Studies* 28.1 (1983): 63–84; and Gad Freudenthal, "Arabic and Latin Cultures as Resources for the Hebrew Translation Movement," in *Science in Medieval Jewish Cultures*, ed. Gad Freudenthal (New York: Cambridge University Press, 2011), 74–105.

15. Isaac ben Solomon Israeli (Egypt, c. 850–950), Sa'adiah ben Yosef Gaon (Egypt, 882/892–942), Abu Sahl Dunaš ibn Tamim (al-Andalus, tenth century), Judah b. David Ḥayyuj (Fez, tenth–eleventh centuries), Baḥya ibn Paquda (Saragossa?, c. 1050–c. 1156), Mošeh ibn Ezra (Granada, c. 1055–after 1138), Yehudah ha-Levi (Tudela, 1080–1141), and Maimonides (Cordova, 1135–1204), among many other Jewish philosophers, scientists, grammarians, poets, and biblical interpreters, used Arabic or Judaeo-Arabic in all or most of their writings.
16. After Ibn Ezra, many Jews would write in Hebrew or would translate into Hebrew treatises on science and philosophy in the Iberian Peninsula and southern France: Samuel ben Judah ibn Tibbon (c. 1160–c. 1232), Moses ibn Tibbon (fl. 1244–1274), Jacob ben Maḵir Ibn Tibbon (ca. 1236–1307), Levi ben Gerson (1288–1344), Emmanuel ben Jacob Bonfils (1300–1377), Jacob ben Abba Mari Anatoli (b. c. 1200), Šem Tov ben Joseph Falaquera (c. 1225–1295), among others. Still, some would also continue to write in Arabic or Judaeo-Arabic, like Maimonides, or Ibn al-Qursunuh (Seville, fourteenth century), who wrote in Arabic on the astrolabe.
17. By "living" I refer to literary Hebrew and do not imply that Hebrew was a language spoken in everyday life.
18. The earliest extant astrolabe that has been dated was made in 927–928 by an Arab named Nastalus or Bastalus (now in the collections of Kuwait National Museum); see David King "An Instrument of Mass Calculation made by Nastulus in Baghdad ca. 900," *Suhayl* 8 (2008): 93–119.
19. For a bibliography of these translations, see Josefina Rodríguez-Arribas, "Medieval Jews and Medieval Astrolabes: Where, Why, How, and What for?," in *Time, Astronomy, and Calendars in the Jewish Tradition*, eds. Sacha Stern and Charles Burnett (Leiden-Boston: Brill, 2013), p. 223, n. 3.
20. Regarding this question, see Charles Burnett, "Hebrew and Latin Astrology in the 12th Century: The Example of the Location of Pain," in *Stars, Spirits, Signs: Astrology 1100–1800. Studies in history and Philosophy of Science Part C: Studies in History and Philosophy of Biological and Biomedical Sciences* 41.2 (2010): 70–75.
21. I have analysed these four treatises in an article, see Rodríguez-Arribas, "Medieval Jews and Medieval Astrolabes". For the chronology of Ibn Ezra's treatises on the astrolabe, see Shlomo Sela and Gad Freudenthal, Sela, Shlomo, and Gad Freudenthal. "Abraham Ibn Ezra's Scholarly Writings: A Chronological List," *Aleph* 6 (2006): 13–55. For an early history of the astrolabe and the treatises written about it, see Otto Neugebauer, "The early history of the astrolabe," *Isis* 40.3 (1949): 240–256. For the critical edition of the Latin version of Ibn Ezra's treatise, see Jose María Millás Vallicrosa, "Un nuevo tratado de

astrolabio de R. Abraham ibn Ezra," *Al-Andalus* 5 (1940): 1–29. Ibn Ezra's treatises on the astrolabe in Hebrew are all in manuscript versions (about fifty manuscripts), and only one of these manuscripts has been published; see Ẓevi Hirsch Edelman (Hebrew ed.), *Abraham ibn Ezra's Keli ha-Nehoshet* (Koenigsberg: [s.n.], 1845).

22. See Norman Golb, *The Jews in Medieval Normandy: A Social and Intellectual History* (New York, Cambridge and Melbourne: Cambridge University Press, 1998), 260.
23. For a list of Ibn Ezra's works in Latin, see R. Smithuis, "Science in Normandy and England under the Angevins: The Creation of Abraham ibn Ezra's Latin Works on Astronomy and Astrology," *Hebrew to Latin, Latin to Hebrew: The Mirroring of Two Cultures in the Age of Humanism*, ed. G. Busi (Berlin-Torino: Nino Aragno Editore, 2006), 23–57.
24. Literally, "the brass or copper instrument" (indicating the metals used to make it) or just "the instrument".
25. This is the Hebrew transcription of the Arabic word for astrolabe (*asṭurlāb* or *aṣṭurlāb*), which is in turn a Greek loan word (*astrolábon*) meaning literally "star catcher".
26. This is a Greek loan word denoting astrology in Rabbinic Hebrew (from the Greek *stegános*, which means "covered, protected", and, eventually, "secret"). It seems that it was also used in medieval Hebrew to refer to the astrolabe (possibly because of its association with astrology and astrologers, who used the astrolabe to cast horoscopes).
27. We find *mešartim* in the second version of Ibn Ezra's *Sefer ha-mivḥarim* (*Book of elections*, Béziers 1148), see Shlomo Sela, *Abraham Ibn Ezra on Elections, Interrogations, and Medical Astrology: A Parallel Hebrew-English Critical Edition of the Book of Elections (3 Versions), the Book of Interrogations (3 Versions); and the Book of the Luminaries* (Leiden-Boston: Brill, 2011), 146 [1.5.3]; and in his commentary on Job 38:7, see Mariano Gómez Aranda [Heb. ed. and Spanish tr.], *El comentario de Abraham ibn Ezra al Libro de Job* (Madrid: CSIC, 2004), 76, (Rome 1140–1142); *maḥberet* in the second version of his *Sefer ha-mivḥarim* (*Book of elections*, Béziers 1148), see Shlomo Sela, *Abraham Ibn Ezra on Elections, Interrogations, and Medical Astrology*, 146 [1.6.6] and in his commentary on Leviticus 23:24, see Asher Weizer [Heb. ed.], *Pirušei ha-Torah le-rabbeinu Abraham ibn Ezra: Wayyiqra', Bemidbar, Devarim* (Jerusalem: Mosad Rav Kook, 1976), 86–87, (Lucca 1142–1145); *ha-qav ha-šaveh* in his *Sefer Mišpeṭei ha-mazzalot*, see Meir Ben Yitzhaq Baqal [Heb. text], *Seder 12 ha-mazzalot*, (Jerusalem: Hotsa'at Bakal, 1994), 2nd v., 166 (Rouen 1154–1157) and in his commentary on Exodus 27:9, see Asher Weizer [Heb. ed.], *Pirušei ha-Torah le-rabbeinu Abraham ibn Ezra, Šemot* (1st and 2nd version) (Jerusalem: Mosad Rav Kook, 1977), 182–183 (Rouen 1155–1157); *gavhut* in his *Sefer ha-moladot* (MS Paris BNF 1056, f. 51a, Béziers 1148) and his commentary on Psalms 46:3, see Menachem Cohen [Hebrew text], *Miqra'ot gedolot ha-keter* (Ramat Gan: Bar Ilan University, 1992), Psalms, v. I, 146 (Rouen 1156); *neṭiyyah* in the first version of his *Sefer ha-'olam*, see Shlomo Sela, *Abraham ibn Ezra's The Book of the World (Sefer ha-'olam)* (Leiden-Boston: Brill, 2009), 60 [14.4] (Béziers 1148), and his commentary on Psalms 19:5, see Cohen, vol. I, 59 (Rouen 1156); *tequfat Nisan* in his *Sefer ha-'ibbur*, see Solomon Z. H. Halberstam [Heb. ed.] (Lyck: L. Silbermann, 1874) 4 (Verona 1146); *ro'š ha-Teli* in the second version of his *Sefer ha-te'amim*, see Sela, *The Book of reasons*, 198 [2.7.9] (Rouen 1154) and his commentary on Joel 3:4, see Uriel Simon [Heb. ed.], *Abraham ibn Ezra's Commentaries on the Minor Prophets, vol. I: Hosea-Joel-Amos* (Ramat Gan: Bar Ilan University, 1989), 163–164 (Rouen 1156?); and *zenav ha-Teli*

in the second version of his *Sefer ha-te'amim, see* Shlomo Sela, *Abraham ibn Ezra's The Book of Reasons (Sefer ha-te'amim)* [Heb. ed. and English tr.] (Leiden-Boston: Brill, 2007), 198 [2.7.9] (Rouen 1154), and his commentary on Exodus 12:2, see Weizer, *Šemot,* 71 (Rouen 1155–1157).

28. We also find *ma'eṣar / ma'eṣor* and *beriaḥ* in the first version of Ibn Ezra's treatise on the astrolabe (Edelman, 103 and 101, respectively). *Sar* and *meqomot ha-ḥayyim* are also in the second version of Ibn Ezra's *Sefer ha-še'elot (Book of questions,* Rouen 1154–1157*)*, see Shlomo Sela, *Abraham Ibn Ezra on Elections, Interrogations, and Medical Astrology*, 354 [9.1]. Bar Hiyya uses the terms *ha-qav ha-šaveh*, *zenav ha-Teli*, and *tequfah* in his *Sefer hešbon mahalakot ha-kokavim*, see Millás Vallicrosa [Heb. ed. and Spanish trans.], *La obra Séfer hesbón mahlekot ha-kokabim (Libro del cálculo de los movimientos de los astros) de R. Abraham bar Hiyya ha-Bargeloní* (Madrid: CSIC, 1959), 21, line 1 [ch. 4], 5, line 17 [introduction], and 6, line 15 [introduction], respectively), and *roš ha-Teli* in his *Sefer megillat ha-megalleh*, see Adolf Poznanski and Julius Guttmann, *Abraham bar Hiyya, Sefer Megillat ha-megalleh*, (Jerusalem: Verein Mekize Nirdamim, 1968), 11 [ch. 1]).
29. Ibn Gabirol uses *qav ḥešev ha-'afuddah* to denote the solar path or ecliptic along which the twelve constellations of the zodiac lie; see Raphael Loewe, *Ibn Gabirol* (London: Peter Halban, 1989), 135.
30. Bar Hiyya also uses *roḥav* and *merḥav* in his *Sefer hešbon mahalekot ha-kokavim* (see Millás Vallicrosa, 31, line 7 [ch. 6] and 27, line 14 [ch. 5], respectively), and *roḥav in* his *Sefer megillat ha-megalleh*; see Poznanski and Guttmann, 148 [ch. 5]).
31. In the first Hebrew version he transliterates a couple of Arabic terms (transliterated in Hebrew as *'al-ẓuhr* and *'al-'aṣr*), but this is only because these terms denote the times of two Muslim prayers (at noon and afternoon, respectively). The Arabic names of the stars and the lunar mansions are also transliterated in Hebrew; these are justified as they are proper names. One of the rare exceptions is the use of the Arabic loan word *al-samt* in the first version to denote azimuth.
32. Second version, MS BNF 1045, f. 188a: אין כח מבין לתרגם בלשון אחרת כי אם בדרך
קשה ואף כי בלשון הקדש שאין אנחנו יודעים ממנה רק הכתוב במקרא ושמות כלי הנחושת רבים ועתה
אזכירים על פי אשר תשיג יד דעת. All the translations used in this study are mine. Cf. the introduction of the first version, MS BNF 1061, f. 148a.
33. See note 26.
34. Third Hebrew version of Ibn Ezra's treatise on the astrolabe, MS Gunzburg 937, f. 2a: כל מבין לשמוע ידע כי לשון הקדש היתה רחבה מכל לשונות הגוים כי היא היתה
הראשונה לכלם ובעבור גלות עם הקדש מאדמת הקדש התערבו בגוים וילמדו לשון עם ועם ונשכחה
לשונם מפיהם ולא נשארת בידם רק ספרי הנביאים. ואשר לא הוצרכו בספריהם לדבר בו לא מצאנוהו
כתוב על כן הקשי לחדש שמות והטעם לתרגם אותם מלשון אל לשון אחרת. השמות זה הספר יקרא
אסטגנינות וקראתיו כלי הנחשת בעבור שיעשו רובם מנחשת. This passage has parallels in the works of several Jewish grammarians prior to Ibn Ezra; for instance, see Santiaga Benavente and Ángel Sáenz-Badillos (Hebrew ed. and Spanish tr.) *Tešubot Talmide Menahem* (Granada: Universidad de Granada, 1986), 20; in Šelomoh ibn Gabirol's *Sefer ha-'Anaq*, see Haim Brody and Jefim Schirmann (Hebrew ed.) *Ibn Gabirol Secular Poems* (Jerusalem: The Schocken Institute, 1974), 170 (verses 38–40 and 44); in Mošeh ibn Ezra's *Kitāb al-Muḥāḍara wa 'l-Mudhākara*, see Montserrat Abumalham [Judaeo-Arabic ed. and Spanish tr.] (Madrid: al-Majlis al-A'la lil-Abḥāth al-'Ilmīyah, Ma'had 'Ilm al-Lughāt, 1985–1986), 2 volumes, v. I, Judaeo-Arabic text, 57; in Yehudah ha-Levi's *Kitāb al Khazari*, which I quote in full: "The Khazar king asks:—Is Hebrew superior to other languages? Do we not see distinctly that the latter are more finished and comprehensive? To which the Jewish Rabbi replies:—It shared the

fate of its bearers, degenerating and dwindling with them. Considered historically and logically, its original form is the noblest. According to tradition it is the language in which God spoke to Adam and Eve, and in which the latter conversed . . . This is supported by the evidence of the Torah . . . The superiority of Hebrew is manifest from the logical point of view if we consider the people who employed it for discourses, particularly at the time when prophecy was rife among them, also for preaching, songs and psalmody. It is conceivable that their rulers such as for instance, Moses, Joshua, David, and Solomon lacked the words to express what they wished, as it is the case with us today, because it is lost to us?" see Hartwig Hirschfeld [English tr.], *Kitab al Khazari* (New York: Schocken, 1964), 109–110. Concerning all these questions and references, see Ángel Sáenz-Badillos, "Philologians and Poets in Search of the Hebrew Language,' in *Languages of Power in Islamic Spain*, ed. Ross Brann (Bethesda: Cornell University, 1997), 49–75.

35. See Carlos del Valle Rodríguez, *Sefer ṣaḥot de A. ibn Ezra* (Salamanca: Universidad Pontificia de Salamanca, 1977), 161 [XV. 32. 3], 159 ([XV. 31. 15], and 161 [XV. 32. 6], respectively.
36. In this passage we should understand that Ibn Ezra is combining grammatical and mystical meanings when he uses the words *adjective* and *noun* in relation to the biblical names of God. Basically a *noun* is an autonomous element in a sentence, while an *adjective* depends on other elements of the sentence. So he places nouns above adjectives in grammatical sentences, but also (by analogical thought) on a cosmological level; the divine names that are nouns are above the divine names that are adjectives, and each of these two groups has different cosmological roles in relation to Creation, as explained in the the passage to which I refer.
37. See del Valle 1977, 231 [XXVI. 67. 10–12], 233 [XXVI. 68. 8–10], 203 [XXI. 53. 10–15], 205 [XXI. 54. 1–2], and [XXII. 54. 13–14]. What follows in the article regarding the cosmological values of these letters has been analysed in detail and in a different context in Josefina Rodríguez Arribas, "Los nombres de Dios: gramática, matemática y teología en Abraham ibn Ezra," in *Maimónides y su época,* ed. Carlos del Valle Rodríguez, Santiago García-Jalón de la Lama and Juan Pedro Monferrer Sala (Madrid: Sociedad Estatal de Conmemoraciones Culturales, 2007), 491–510.
38. See Weizer, *Šemot*, 27.
39. See Weizer, *Šemot*, 76.
40. For an overview of the concept of language in Ibn Ezra and other Jewish writers of his time, see Irene E. Zwiep, "Did Adam Speak Aramaic? Abraham ibn Ezra's Views on the Origin of Language," *Beiträge zur Geschichte der Sprachwissenschaft* 3 (1993): 45–60; Angel Sáenz-Badillos, "Philologians and Poets in Search of the Hebrew Language"; "The Origin of the Language and Linguistic Pluralism according to Medieval Jewish Exegetes," in *Verbum et Calamus: Semitic and Related Studies in Honour of the Sixtieth Birthday of Professor Tapani Harviainen*, eds. Hannu Juusola et al. (Helsinki: The Finnish Oriental Society, 2004), 293–303; and the general study of Irene E. Zwiep, *Mother of Reason and Revelation, A Short History of Medieval Jewish Linguistic Thought* (Amsterdam: J. C. Gieben Publisher, 1997). According to Zwiep (*ibid.*, 155–156), Ibn Ezra is among those Jews who adopted an intermediary position with respect to the theories about the conventional and the natural (created) origin of language: "If anything, Ibn Ezra's account of the origin of language is a tale of close co-operation between God and man. God's contribution to the development of language consisted in 'being the beginning of all His creations', i.e. in creating man, in creating the animals, and in endowing man with both the need and the faculty to name these things". See also Gad Freudenthal, "Dieu parle-t-il hébreu?," *Les cahiers du judaïsme*

23 (2008): 4–18. For a different perspective on Ibn Ezra's understanding of the Hebrew language, see Luba R. Charlap, "Abraham ibn Ezra's Viewpoint Regarding the Hebrew Language and the Biblical Text in the Context of Medieval Environment," *Folia Linguistica Historica* 26.1–2 (2005): 1–11.

41. For an overview of the principles of expansion of the Hebrew language during the medieval period, see Chomsky, "The Growth of Hebrew during the Middle Ages," *The Jewish Quarterly Review* 57 (1967): 125–133.
42. See Poznanski and Guttmann, Abraham bar Hiyya, *Sefer megillat ha-megalleh*, 113, and Millás Vallicrosa, *Sefer ḥešbon mahalek̲ot ha-kok̲avim*, 19 [Hebrew text].
43. See Poznanski and Guttmann, Abraham bar Hiyya, *Sefer megillat ha-megalleh*, 128 and 149; Millás Vallicrosa, *Sefer ḥešbon mahalek̲ot ha-kok̲avim*, 44 [Hebrew text]; Edelman, *Abraham ibn Ezra's Keli ha-Nehoshet*, 102.
44. As Guttmann explains in the introduction to Bar Hiyya's *Sefer megillat ha-megalleh*, the form most frequently used in the manuscripts to denote angle is *yated*, despite the frequent use of *yeter* in some manuscripts; see Millás Vallicrosa, *Abraam bar Hiia, Llibre revelador* (Barcelona: Editorial Alpha, 1929), 48–49.
45. See note 27.
46. For an analysis of this matter in some of Ibn Ezra's biblical commentaries, see Josefina Rodríguez Arribas, "Astronomical and Astrological Terms in Ibn Ezra's Biblical Commentaries: A New Approach," *Culture and Cosmos* 13.1 (2009): 3–21. The subsequent lines in this paper are taken from this article.
47. Norman Roth, "The 'theft of philosophy' by the Greeks from the Jews," *Classical Folia* 32 (1978): 53–67. See also Shlomo Sela, *Abraham ibn Ezra and the Rise of Medieval Hebrew Science*, 104–6 and 140–43.
48. Indeed, Ibn Ezra states in his long commentary on Genesis that the multiplicity of languages appeared as a consequence of the people (Israel) scattering from the place (the land of Israel) and forgetting their first language (Hebrew); see Sáenz-Badillos, "The Origin of Language," 301. Concerning the exile as the main reason for the decay of the Hebrew language, see Zwiep, *Mother of Reason and Revelation,* 229–40.
49. Raymond P. Scheindlin, "Hebrew Poetry in Medieval Iberia," in *Convivencia: Jews, Muslims, and Christians in Medieval Spain,* eds. Vivian B. Mann, Thomas F. Glick, and Jerrilynn D. Dodds (New York: W. W. Norton & Company Inc, 1992), 44.
50. As a comparative insight into the kind of problems that the Greeks faced in finding technical terms in for philosophical concepts, see E. A. Havelock, "The Linguistic Task of the Presocratics," in *Language and Thought in Early Greek Philosophy,* ed. Kevin Robb (La Salle, IL: Hegeler Institute, 1983), 7–81.
51. On a planispheric astrolabe there are several of these lines starting from the horizon, which is the first almucantar of an astrolabe. Thus almucantars are circles (or segments of circles on a planispheric astrolabe) that indicate different altitudes of stars for a specific latitude. Southern astrolabes would have these lines engraved in the lower half of the instrument.
52. See Rika Gyselen, Anna Caiozzo, et al., *La science des cieux: sages, mages, astrologues* (Bures-sur-Yvette: Groupe pour l'étude de la civilisation du Moyen-Orient, 1999), 25 and 31, and Anna Caiozzo, *Images du ciel d'orient au moyen âge:une histoire du zodiaque et de ses représentations dans les manuscrits du Proche-Orient musulman* (Paris: Presses Paris Sorbonne, 2003), 213–225.
53. See http://www.britishmuseum.org/explore/highlights/highlight_objects/pe/a/hebrew_astrolabe.aspx (copyright The Trustees of the British Museum, 2011, accessed March 7, 2015).
54. Here we will take components of the astrolabe to be those that are apparent in the body of the astrolabe, whether removable pieces or elements inscribed in the metal.

8 *Opyt* in the Social Lexicon of Modernity

The Experience/Experiment Dichotomy

Ekaterina Smirnova

> *Experience is so much a part of everyday language, so imbricated in our narratives that it seems futile to argue for its expulsion . . . Given the ubiquity of the term, it seems more useful to work with it, to analyze its operations and to redefine its meanings*[1]

1. INTRODUCTION

In Russian the doublet experience/experiment is expressed by the word *opyt*, which describes complete knowledge and skills (*experience*, *expérience*, *Erlebnis*), a specially conducted and controlled procedure of testing/examination (*experiment*, *l'expérimentation*, *Erfahrung/Versuch*), or a literary probe (*essay*, *essai*, *Essai*).[2] The period of modernity in Russia at the beginning of the eighteenth century was marked by a renaissance in the spiritual sphere as well as by the development of industry in light of the results being achieved by the exact sciences. The institutionalization of experimental science in Russia was accompanied by the adoption of Western scientific terminology mostly originating from Latin. In the earliest scientific books translated into Russian, different Latin terms (such as *natura*, *phaenomenon*, *observatio*, *experientia*, *experimentum*) were widely used.[3] Up until the mid-nineteenth century, Russian scientists corresponded with their colleagues (even in Russia) in Latin, French, or German. However, the necessity for popularization of the natural sciences (permanently developing alongside industrialization) laid the basis for the design of a national scientific terminology. This was most commonly achieved through direct replication. Later on, the first three concepts were replaced by the Russian equivalents *priroda* (nature), *yavlenie* (phenomenon), and *nablyudenie* (observation). *Experientia* became obsolete. Subsequently the Russian word *opyt* was not only imported from the West but started to be used with the same meaning as the foreign analogue *experiment*.

In the period of modernity the concept of *time*, however, acquired a different connotation which was not the case in the Middle Ages. It moved away from indicating *time of God to the new meaning of time of the individual.*[4]

As a consequence of this reconsideration of time in a historical context, a new active subject emerged, who, based on a series of feelings and perceptions, discovers and experiences the world around. This chapter offers an analysis of the history of the Russian word *opyt* as a mirror reflecting the dynamics of culture and processes, which determined, to a certain degree, contemporary views and ideas through transformations of semantic meanings. Taking into account the interconnection of the meanings of the word *opyt* within a European context, we will examine its conceptualization in Western thought.

2. A HISTORY OF EXPERIENCE/EXPERIMENT IN WESTERN THOUGHT

In practically all works devoted to a study of the concept *experience* we find conclusions which underline its contradictory character. This is most succinctly captured in Michael Oakeshott's assertion that among all the terms dictionary, the notion of experience is very much difficult to handle,[5] and furthermore, "however paradoxical it may sound, the concept of experience is one of the most obscure philosophical terms".[6] This ambiguity is seen even at the lexicographical level: "It permits a distinction between the noun 'experience' as something, that one can be said to 'have' or 'to have learned from' and the verb 'to experience' or the process 'experiencing', the latter suggesting that one is now 'doing' or 'feeling'".[7] The same can be said of *experiment*, which, according to David Hume, implies "the interference of the one who observes phenomenon and governs them in the hope of a better understanding of their mechanisms".[8] *Experience* in a broader sense of the word means that we find a limit which allows us to discover and explain something. Often it refers to the things or phenomena which seem to be reluctant to change or are conceived of as hard to be shared with others as, for example, religious experience; "personal and incommunicable; however abundant the assurance of Reality it brings to the experiment, it cannot share the certainty with others as the normal experience of the senses does".[9]

> *Experiment*, however, is a test adopted with uncertainty as to whether it will answer the purpose. "An Observation or Experiment carefully made . . . leads us with greater Certainty to the Solution".[10] It is an operation carried out in order to discover something, to ascertain or illustrate some truth. It always implies a tangible manipulation, in the same way as experiments in physics do: "If the experiment is performed in an exhausted receiver . . . the substance forming the electrodes is condensed".[11] So *experiment* is performed; *experience* is accumulated on the basis of a series of observations made, upon which we make our decisions. The "only ultimate source of our knowledge of nature and its

> laws, experience; by which we mean . . . the accumulated experience of all mankind in all ages, registered in books or recorded by tradition".[12]

In many cases both terms go together and one implies another: "A regular *series of experiments* can provide a ground for the appearance of *experience*. Or in a contradictory sense, *experience* as past experience becomes autonomous and, as such, becomes one in a series of experiments".[13] In his work Hans-George Gadamer contrasts experience with experiment as an element of scientific experimental method. In one of his interviews he defined the interconnection between these two terms in the following manner:

> To have an experience does not mean to know something once and for all and to stay firm with that knowledge. An individual possessing an experience is not a dogmatist. Experience provides freedom to gain new experience and to find out new things. I refer to it to an endlessness of experience.[14]

Despite the fact that "the authority derives from experience",[15] it highlights only what has been, but never what must be. The history of this concept and its analysis in different research traditions have been a focal point for many researchers. The first thoughts on the subject can be found in Greek philosophy.[16]

Aristotle distinguished two ways of learning—observation and reasoning (*empeiria*). The initial form of learning was an *empeiria* based on the senses, feelings, and perceptions of a subject. The famous formula created by Aristotle was: "nihil in intellectu quod non prius in sensu" (nothing is in the intellect, which was not first in senses). In his work *Posterior Analytics* Aristotle wrote that "from perception based on the senses emerges memory; repeated memories give rise to experience [*empeiria*]". The interpreters of Aristotle's writings believed that experience implied for Aristotle "a habit which allows an individual to act in similar situations meaning in those where he already had some recollections, some similar moments of information registration". Thus it involves guess-belief formed on the basis of a series of perception and memory.[17] From *empeiria* arises a principle of craft (it is about what comes to be) or a principle of science (it is about what is).[18]

In the thirteenth century it was the Franciscan friar Roger Bacon (1214–1294), an ardent proponent and practitioner of the experimental method, who paid particular attention to the distinctions between *experience* and *experiment*. Experience in his interpretation is double-faced. He makes a distinction between the inner spiritual experience and the external one gained by means of the senses. It is this form of experience which lies at the heart of experimental science and scientific research. He was the first person to use the expression "experimental science" (*scientia experimentalis*) in his work *Opus Majus*. According to Bacon, this new experimental science is superior to all other forms of knowledge thanks to three prerogatives: the first being that it helps to establish credible knowledge, the second refers to its capacity to acquire true knowledge in those fields where other sciences are useless, and the third distinctive feature of experimental science concerns its ability to uncover the secrets of Nature. The ideas of Roger Bacon were later developed by Francis Bacon (1561–1626). At this time, reliability, credibility, and replication became important factors in the realm of the production and circulation of knowledge: "The right method could be learned by others; experience must be communicable, not merely intelligible". This new science was based on doing rather than on mere reasoning.

The scientific revolution of Nicolaus Copernicus (1473–1543), Galileo Galilei (1564–1642), and Isaac Newton (1642–1727), as well as further achievements in the physico-chemical sciences in the sixteenth and seventeenth centuries, led to the complete transformation of science. In this matter scientific experiment, mathematical methods, and techniques of measurement played a decisive role. Bacon described this new science as such: "Performing an experiment it is necessary to be precise with the methods used in order that the observers and readers could judge whether it is true or false to be able to produce more exact experimental techniques in future".[19] In the eighteenth century the basic principles of this new science were adopted by Mikhail Lomonosov (1711–1765) for the Russian Academy of Sciences.[20]

Before the eighteenth century the word *opyt* and some of its derivatives existed in the Russian lexicography. For the purposes of this study it is necessary to look at their usage in Russian historical documents. The term *opyt* first appears in the sixteenth century as a derivative of the verb *opytati*, used in a juridical context when the matter regarded the procedure of investigation. This is an article from Juridical Codex written at the end of the fifteenth century under the reign of the Russian tsar Ivan III:

> It is necessary to *opytati* [to interrogate] a person who is being slandered. If that person has already been accused of any kind of crime, he is subjected to torture; if not—the questioning of others [i.e., possible witnesses] should first be undertaken.[21]

According to this article, evidences provided by the accused party contra somebody else is the subject of a detailed investigation. If the person who is accused has already been implicated in crimes he immediately becomes the subject of questioning; if he is not a subject to any testimonies pointing to a previous offence, he is bailed until the questioning of witnesses has been carried out in order to find out the reputation of the accused party. Later, in the mid-sixteenth century, the term *opyt* was already used in the sphere of trade when matter concerned the examination of the goods quality:

> Under the reign of your father [Gustav I] our merchants of fat and wax went to Sweden, and your father *pytkoi pytal* [scrutinized by *opyt*, i.e., tested/examined] all the goods by himself behaving like just a common man.[22]

This is an extract from the message of the Russian tsar Ivan IV to the Swedish king Johan III in which he blames his Swedish counterpart for carrying out an appraisal of these Russian goods manually. In Ivan's opinion it should not be a matter for a noble to do this. He accuses Johan of being of dishonorable origin.

In the mid-seventeenth century *opytati* with the meaning of undertaking tests enters the industrial sphere, where *opyt* takes on the meaning of a sample/object of examination. In this example, dating to 1645, the matter is an examination of the quality of ores found in the Urals. The text contains instructions, for the miners, on how to test the qualities of minerals and orders them to send the samples to the tsar, in Moscow, for final verdict:

> The voivode, Maksim Streshnev, delivers an order to search for copper ore in Siberia. After it is found the copper ore should be sent to Moscow and *opytyvana* [examined] by state officials. Once examined it is discovered that the *opyts* [samples] presented, contain no copper at all. Consequently another order is delivered to find a true sample.[23]

At the beginning of the eighteenth century *opyt* as a testing procedure is used in the case of the evaluation of military equipment.

> We should accept for practical usage any gun only by *opyt* [examination/test] and by true and credible evidence.[24]

In this passage from his work "Reflections on Military Behavior", Ivan Pososhkov, an advocate of the reforms carried out by Peter the Great, warns that in the industry of gunpowder production only Russians should be employed since it could be strategically hazardous to put such a paramount sector into a foreigner's hands. What is important here is the appearance of credible evidence—the all-important element of scientific practice. After the founding of the Russian Academy of Sciences and the emergence of the exact sciences, *opyt* with the meaning of test/examination (*opyt* as experiment) is incorporated into scientific practices.

3. INSTITUTIONALIZING THE PRACTICE OF SCIENTIFIC EXPERIMENT IN RUSSIA

A milestone in the development of science in Russia was the foundation of the Academy of Sciences. Its initial establishment began in the early years of Peter the Great's reign when he began to negotiate with prominent Western scientists. Peter's inspiration for an Academy of Sciences emerged during his grand tour of Europe (1697–1698), when the future emperor visited Germany, the Netherlands, England, and France, and he met with well-known European savants and philosophers, among them Gottfried

Wilhelm Leibniz and Isaac Newton. In 1712 in a letter to the Russian tsar Leibniz wrote:

> You can . . . import the best and improve by the appropriate measures much of what has already been achieved in both parts of the world; since in your state all which concerns the development of science is new and resembles a sheet of white paper, and consequently you can avoid many of the mistakes which Europe underwent gradually so far.[25]

On January 13, 1724, Peter the Great signed the so-called "Definition on the Academy" in the Senate. The document stated that "Henceforward the goal of the development of different *hudezhestva* [arts/practical knowledge] is pursued in Russia". This meant science was no longer regarded as a sphere of pure curiosity but was rather perceived as a domain of practical decision-making. The Russian Academy of Sciences "had to all the découvert(e)s [discoveries] discuss and present their approval [approbation] . . . 1. Are they reliable? 2. To what extent they are useful? 3. Had something of that kind been already invented before?"[26]

Theoretical and experimental physics were introduced as separate disciplines in the Academy from the first days of its existence. Its charter contained the following regulations:

> In the second class there is physics theoretical and experimental. Experimental examination of the physical theories and correlation of speculations with experiments (in other words observations with *opyts*) was regarded as an obligatory practice in physics.[27]

When the idea of independent research was established in Russia, the first step in such a direction was the *Kunstkamera* (Chamber of Curiosities) collection, the first museum of historical and natural artifacts in Russia, employing various instruments and mechanisms as an experimental basis for the future development of science.

In the mid-eighteenth century, with the efforts of Georg Kraft and Georg Wilhelm Richmann, the physics cabinet was founded in the Academy, which contained more than 350 various instruments. They were classified into the following categories: "Mechanics", "Optics", "Magnetism", "Warmth and Meteorology"; "The category of 'Mechanics' contained about 180 various instruments: the majority were made by Leutmann Johann Georg, which included mechanical clocks, air and water pumps, different cylinders and tubes".[28] Among the optical instruments there were some binoculars, telescopes, three cameras obscura, "an artificial eye incrusted in ivory", "an incendiary mirror, grand metallic rotating", made by Jacob Bruce in 1735, different microscopes invented by Pieter van Musschenbroek, mirrors of different forms—inverted, canonical, round, plane and complex. In the section "Magnetism" there were magnets of different force, compasses, and sundials. In the section "Warmth and Meteorology" were included thermometers and barometers made by the best foreign craftsmen as well as various meteorological instruments made in the Instrumental Chamber of the Academy of Sciences by such academicians as Leutmann, Kraft, and De L'Isle.[29]

In December 1775 the Commission governing the Academy of Sciences Affaires took the decision to offer public lectures. Lectures in physics were to be read by Georg Kraft. At first it was decided to hold them in his apartment, where "necessary instruments were available since they were to be accompanied by *opyts*".[30] "Physics began to collect various observations and opyts in nature, compare them and search for their possible use; this contributed to the progress in physics and it became to be a science with solid ground", wrote Feodor Politkovski in his work *A Word on the Connection of the Natural History with Physics, Chemistry and Medicine*.[31]

Before the eighteenth century in Russia there was no experimental science, and the meaning *scientific experiment* was not assigned to the word *opyt*, though it could signify, as we showed prior, an examination including elements of simple research. From the second decade of the eighteenth century, the word *experiment* can be found in different Russian treatises: "We should think about inviting from England such a scientist who could conduct various experiments and make necessary instruments".[32] Or another example, "he needs a special machine for demonstrating experiments he does with fire and light".[33] In the educational program *Rules of Instruction* (1728) it was proclaimed to become a necessary condition of any lecture: "Everything should be confirmed by certain models and experiments".[34]

These words can be found in such publications as *Primechanija na Vedomosti* (Additions to Vedomosti):[35] "We can make a serious mistake in trying to understand and explain the causes of this or that phenomenon before it is confirmed by a series of observations and experiments".[36] However, Mikhail Lomonosov (one of the architects of the new emerging scientific community and a founder of the Russian scientific school) preferred to use the Russian equivalent *opyt* instead of the Latinism *experiment* even in his translations taken from Latin:

> In modern times the sciences have developed so greatly that those who lived a thousand years earlier would hardly recognize them. This stems from the fact that scientists, investigators of natural things, pay less attention to the ideas appearing in somebody's head but instead are convinced by a trustworthy art [a series of *opyt*-experiments]. The main element of the natural science, physics, takes this principle as its basis. Reliable arguments can only be produced from a series of *opyts*.[37]

The whole point of this meaning of the term *opyt* is to reproduce a phenomenon in artificially created and controlled conditions in order that it may be reproduced. Something is proved when being observed a certain number of times:

> 1757 There are evidences provided by the individuals possessing credibility which indicate that by repeated melting silver can be turned to gold. These and other *opyts* would prove this if they could be observed.[38]

A retrospective analysis of the practices described as *opyt* leads then to the following conclusions:

(1) the ancient practice of questioning/interrogation and investigation (particularly in judicial and clerical contexts) was expressed in Russian in the sixteenth century by the verb *opytati*;
(2) the practice of *opyt* as an expertise used in the fields of agricultural and industrial examination of the quality of goods beginning in the seventeenth century was transferred to an expertise of military equipment in the late seventeenth and early eighteenth centuries.
(3) the practice of scientific *opyt* [experiment] appeared in the eighteenth century due to the institutionalization of natural sciences in the Russian Empire modernized by Peter the Great, and the term *opyt*, meaning examination/verification, was transferred from the spheres of trade and industry (sectors) into the field of exact sciences.

A series of *opyts-experiments*, carried out by different individuals, is aimed at proving something. It is a practice of establishing credibility.

4. ART AS A SCIENTIFIC PRACTICE

Besides the term *opyt* for description of scientific practices in the eighteenth century, the following words were also used: *iskus* (trial, test, ordeal), *iskushenie* (temptation), and *iskusstvo* (art in a sense of practice), all of which describe the process of examination/observation. They meant practical skills undertaken in order to achieve the end product as the result of knowledge and practice. The following are some examples of their usage:

> Air can be both cold and warm, a fact which is confirmed by *iskusstvo* [everyday art/practice].[39]
>
> Everyday *iskusstvo* [art] confirms that water can flow without any troubles from the high areas to low ones.[40]
>
> Through *iskus* [*opyt*] we know that after the assent the solid D is immovable and the solid A moves with the same speed as did the solid D did before the assent.[41]
>
> A chemist should practice science not only by readings but by his own *iskusstvo* [art] conducting a series of opyts [experiments].[42]

> Even the very mistakes of the ancestors are valuable since without them it would often be impossible to *iskushat'* [investigate] the nature not having ever undertaken the risk of being mistaken.[43]

In his letter to Ivan Shuvalov,[44] Mikhail Lomonosov orders work to be carried out on a first Russian textbook on obstetrics:

> It is necessary to choose a number of books on the *iskusstvo* [art] of obstetrics and take the best one in order to reproduce the same in Russian. As supplementary information it is necessary to describe the skills and abilities which the Russian *iskussnye* [experienced] midwives have attained and make suggestions for its substitution.[45]

In scientific *opyts*, the art (*iskusstvo*/skills) of a researcher is focused on the reproduction of a phenomenon in the laboratory conditions close to those we observe in natural settings. In order to distinguish the former from the latter, the first were called *iskustvennyi*. The emergence of this word dates back to the mid-eighteenth century: "In production of chemical operations it is necessary to purify thoroughly the objects produced in natural and artificial [*iskustvennoi*] manner which are the subject of chemical research";[46] "We should say a few words on the inner characteristics of the mixed bodies which can be both natural and artificial . . . so to say existing both in nature or having been specially produced".[47]

In the texts by Lomonosov's works, those who undertook experiments were defined by the adjective *iskussnyi*,[48] and in European languages names for these specialists were distinguished by the root *exper*: *expérimenté* (French), *experimental* (English). It is possible that "*isskusnyi* chemist" (the one who made a series of experiments) was transformed into "*opytnyi* chemist" under the influence of the English expression "experimental chemist" or its French analogue "chimiste expérimenté".[49]

In the European languages in the eighteenth century the practice of gaining knowledge based on a series of examinations/observations was assigned to the term *experientia* and not to the term *art*. There were even some efforts to introduce this word into Russian lexicon. However, these attempts were not adopted: "Experientia, which everybody refers to is not commonly accepted"[50] or "it becomes clear through expertientia".[51] Sometimes total confusion reigned in the usage of the words: "The main characteristics of water we learn either by *povsednevnoe iskusstvo* [everyday art] or by conducting particular physical experiments".[52] This example is illustrative in that it demonstrates the dualism of the Russian concept *opyt* as experiment and as a practice of a series of observations/examinations leading to the accumulation of knowledge-experience. The words *iskus* and *iskushenie* retain only the meaning of experience lived through and are completely removed from the scientific field, and *iskusstvo* (apart from the creative work)[53] holds only the meaning of possession of certain skills and has lost its meaning as an element of research practice.

5. *OPYT*/EXPERIMENT IN CONTEMPORARY USAGE

Linguistic tradition, gradually formed in the Russian scientific community, is rather peculiar. In contemporary scientific vocabulary scientists use both the terms with equal frequency.[54] According to dictionaries of contemporary Russian language, the practices covered by both terms are always similar if not completely identical. In the *Dictionary of Contemporary Russian Literary Language* (Moscow, 1965) we find that *experiment* is a scientifically conducted *opyt*, and *opyt* is testing of the scientific knowledge in special laboratory conditions. The *Dictionary of the Russian Language* (Moscow, 1981–1984) argues that both *experiment* and *opyt* concern the reproduction of any phenomenon in artificially created conditions with the aim of exploring something new. However, a more detailed analysis of the current usage of *experiment* and *opyt* allows us to reveal some nuances, presented in Table 8.1.

Table 8.1 Stable phrases with the words *opyt* and *experiment* in a contemporary Russian context (in their discussions physicists make these distinctions)

Opyt	*Experiment*
A. When any significant experimental discovery is given the name of its author, it is always referred to as *opyt*. Example: the experiments by Torricelli (atmospheric pressure measurement) are referred to in Russian translations as *opyt*. The same can be said about Pavlov's *opyts* in his study of conditioned reflex.	A. The sections of physics are usually divided into theoretical and experimental parts. In the times of M. Lomonosov there was an expression *opytnaja fisika* (experimental physics), but nowadays it sounds like an anachronism—even in the *Dictionary of the Church Slavonic Language* published in 1847 we read *eksperimentalnaia fizika* (experimental physics)
B. *Opyt* is an activity with a predictable and well-planned result. It is conducted in artificially created conditions. It either confirms or denies some *a priori* ideas. That is why we usually speak about series of conducted *opyts*. Such accounts usually have a three tiered structure: preparation/realization/interpretation.	B. On the contrary, experiment refers to an activity with unclear and unpredictable results. It is often related to a search for something new.
C. The term *opyt* is widely used in the field of natural sciences in which experimentation inevitably involves a host of simple manipulations. For example, in the context of education *opyt* is a demonstration of a phenomenon to an auditorium.	C. The term *experiment* dominates in the fields of modern experimental physics relating to complicated technical equipment (reactors, accelerators, etc).

6. FROM *OPYT*-EXPERIMENT TO *OPYT*-EXPERIENCE

We have already noted that profound semantic changes of many Russian words, which took place in the eighteenth century, were linked to the cultural revolution of the Enlightenment. Earlier we analyzed the process by which the term *opyt* gained the meaning of a *scientific experiment*. Further we will look at the way in which *opyt* took on the meaning of *experience*. The examples which follow demonstrate that this semantic novelty became established in the same historical period and that its emergence is directly linked with the appearance of the term *opyt* as scientific experiment. In the first instance, the word *opyt* means a concrete procedure, demonstrating some kind of phenomenon. In texts written by Lomonosov we find different examples of this kind: "It is necessary to pay attention to the *opyts* made in other places, the success of which has been under question so far";[55] "In many theses defended in practical sciences we often find the results of new *opyts* and observations".[56] Later there emerged references to *opyts*, in general, as a particular method for scientific inquiry. In this case the plural form is often used, "opyts performed confirm the following theories",[57] or, another example, "this contradicts to all opyts".[58] In these examples *opyt* implies experiments-manipulations, but, at the same time, we also observe here a tendency for the emergence of a new meaning: *opyt* as accumulated knowledge. This tendency becomes clear when the plural form is being replaced by a singular one: "*Opyt* demonstrates that there exist several different kinds of water".[59] What concrete experiment demonstrates is not an appropriate question. This is illustrated by the whole set of *opyts* performed as *accumulated experience*. A fine demonstration of the emergence of this new term is the following phrase: "It is quite clear that this contradicts firstly to *opyt* and then to common sense".[60]

Gradually *opyt* emerged in spheres beyond the natural sciences. In Lomonosov texts such examples we find, first of all, in administrative documents (reports, petitions, and memorials): "It is necessary to be aware of and struggle against what was regarded as dangerous for the success of sciences under the ancient and modern state of affairs in the Academy";[61] "It becomes clear through *opyts*, that many efforts were undertaken for the invitation of foreign members".[62] And finally the notion of everyday experience appears: "It becomes clear from mechanical laws and is confirmed by everyday experience that the smaller and lighter an object is, the less it is opposed to the motive forces, and consequently the less is driving force";[63] "Nobody can obtain a post as ordinary academic without demonstration to the scientific community the *opyts* of his own art [skills]".[64] Gradually, in the contexts of sciences, the mentioning of *iskusstvo* disappears and the adjective *opytnyi* as possession of *opyt*-experience comes into the picture: "It is quite clear to those experienced in chemistry, which different effects can be observed".[65]

Another factor favoring the term *opyt* gaining the meaning of experience was its use in the titles of scientific and philosophical works which in

European languages are expressed by such terms as *essay*, *treatise*, *Abhandlung*. It was widespread in the translations where *opyt* was used as the equivalent of essay.[66] This favored for the extension of the semantic field: the word *opyt* lost its connection with the procedure of research-manipulation and gained the meaning of philosophical analysis, directed towards obtaining, extension, and precision of *opyt*-experience.

7. CONCLUSION

According to the famous expression by Reinhart Koselleck (1923–2006), "a concept is not only an indicator of the relations it describes, but also their factors. Every concept establishes certain horizons and limits of possible experience and theory".[67] The application of this principle to the history of Russian science provokes an unintended effect. In particular, it becomes clear that many of the scientific terms adopted in the eighteenth century were later replaced by Russian equivalents. One of the factors helping us understand such fast "assimilation" of European scientific traditions by the emerging Russian sciences were trade and industry expertise which had actively developed in Russia before the reforms of Peter the Great.

At the time when natural sciences were transported from the West to Russia (during the reign of Peter the Great) the practices similar to the *experiment* and denoted by the term *opyt* were already widely used in various discourses—clerical, judicial, trade, and industrial systems. The adherents of natural sciences (mainly imported from the West) could not ignore this fact. However, the frames of practices covered by the Russian term *opyt* were much larger than the procedures denoted by *experiment*. This may explain the parallel usage of both terms *opyt* and *experiment*. Another explanation for the persistence of the rather vague concept *opyt* in the Russian lexicon of exact sciences can be drawn from an epistemological analysis. Historically, one of the meanings of the Russian term *opyt* denoted the final verdict for truth. This notion corresponded to the basic idea of positivism shared by pioneers of natural sciences.

In contemporary Russian language, *opyt* as sample/probe became obsolete. It retained only the meanings of the process of investigation itself and its results. Knowledge gained during the process of *opyt* (*experiment*) extends the *opyt* (*experience*, *Erfahrenheit*): "It allows us both to 'appeal' to experience as it were always a thing in the past and to 'hunger' for it as if it were something that one may enjoy in the future".[68] This dialectic dualism between *experiment* and *experience* is reflected in the Russian language in which both these categories are expressed by the same word *opyt*. This polysemy (combining in one word both scientific procedure and skills gained) contributed to the birth of the term *opyt* as the materialistic philosophical concept of the final criteria of truth.

NOTES

1. Joan Scott, *The Evidence of Experience* in *Critical Inquiry*, 1991, 17.4, p. 774. The most significant work devoted to the concept *experience* is Martin Jay's book *Songs of Experience* (Berkeley and Los Angeles: University of California Press, 2005), in which the Western-European discourse, from the sixteenth to the twentieth century, is presented and multiple meanings of the term are analyzed. Particular attention is given to an analysis of different contexts of its usage (epistemic, religious, esthetic, political, and historical) and to the role of *experience* in different theoretical approaches (American Pragmatism, British Marxism, the works of the Frankfurt School and French Post structuralism), as well as in the writings of such philosophers as Montaigne, Bacon, Lock, Hume, Kant Oakeshott, Collingwood, and Ankersmit.
2. *Experience* in Russian transmits the meanings of feeling and temptation, *experiment* has the meanings of test, examination, and probe. *Essay* is just an essay.
3. Ljudmila Kutina, *Formirovanie terminologii fiziki v Rossii v 1/3 XVIII veka* (Formation of terminology in physics in the 1/3 of the eighteenth century) (Moscow–Leningrad: Nauka, 1966). *Materialy dlja istorii Akademii nauk* (Materials for the History of the Academy of Sciences). Vol. I (1716–1730) (St. Petersburg, 1885). Galina Smagina, *Akademija nauk i rossiskaia shkola vtoroi poloviny XVIII veka* (Academy of Sciences and Russian school in the second part of the eighteenth century) (Moscow: Nauka, 1986).
4. Viktor Zhivov, "Istoria ponjatii, istoria kul'tury, istorija obschcstva" (History of concepts, history of culture, history of society) in *Ocherki istoricheskoi semantiki russkogo yasyka rannego Novogo vremeni* (Essays on Historical Semantics of Russian Language in Early Modernity) (Moscow: Yazyki slavjanskih kul'tur, 2009), 28.
5. Michael Oakeshott, *Experience and Its Modes* (Cambridge: Cambridge University Press, 1933), 9.
6. Hans-Georg Gadamer, *Truth and Method* (New York: Bloomsbury Academic, 1986), 310.
7. Martin Jay, *Songs of Experience. Modern American and European Variations on a Universal Theme* (Berkeley and Los Angeles: University of California Press, 2005), 12.
8. David Hume, *A Treatise of Human Nature* (Oxford: Oxford University Press, 2000), 333.
9. Edward Grubb, *The Religion of Experience: An Examination of Some of the Difficulties of Christian Faith* (London: Nabu Press, 1918).
10. James Keill, *Essays on Several Parts of the Animal Oeconomy* (London: George Strahan, 1717), 417.
11. William R. Grove, *Contributions to Science. On the Correlation of Physical Forces*, 1850, 50.
12. John F. Herschel, *A Preliminary Discourse on the Study of Natural Philosophy* (Philadelphia: Carey and Lea, 1831), 76.
13. Hume, *Treatise*, 142.
14. Hans-Georg Gadamer, *Gadamer and Conversation: Reflections and Commentary*. Ed. and trans. by Richard E. Palmer (New Heaven: Yale University Press, 2001), 52–53.
15. Richard Whately, *Elements of Rhetoric* (London, 1828), 7.
16. The Latin term *experientia* also means as well *essai*, *epreuve*, *pratique* and has the same root as the word *periculum*—meaning *epreuve*, *risqué*—and *peritus*, meaning *habile*, *expert*. In more general terms the root *per* means an advancement (*aller de l'avant*) or penetration. This word means at the same time a

breakthrough and an advancement in the inside and outside world and a gaining of skills and qualification based on experience (*acquis*). *Vocabulaire européen des philosophies*. Sous la Direction de Barbara Cassin, (Paris, Le Seuil/Le Robert, 2004), 436.

17. Patrick H. Byrne, *Analysis and Science in Aristotle* (Albany: SUNY Press, 1997), 175. Travis Butler, "Empeiria in Aristotle" in *Southern Journal of Philosophy* 41, no. 3 (2003): 332.
18. Byrne, *Analysis,* 347.
19. Experiment was often associated with so-called *low sciences* such as alchemy and astrology, and the term «*Emperick*» was used as a pejorative (often in medical discourse) as a synonym of charlatanism until the eighteenth century. In the book *Translations from the Encyclopedia* composed in Russia in the eighteenth century we read the following in an article devoted to the word *Charlatan*. "It is a history of the most prominent Empiricks who cheated the court in France", and, further, in the article entitled *Dogmatique* we find the following passage: "The important difference between two kinds of sects (Empiricks and Dogmatiques respectively) consisted of the fact that Dogmatiques tried to work through the symptoms) observed to get a better understanding of the necessary remedies; whereas Empiricks (didn't pay much attention to this kind of examination and their ultimate exercise was just to make up prescription based on practice" (p. 54). Such a view has some origin in the writings of the Ancient Greek philosophers. For example, in the *Laws* Plato distinguishes between the medical treatment of slaves and that of citizens In the former case, the doctors make practical recommendations immediately, relying purely on their *empeiria* (a series of memories); they have no interest in discovering the account (logos) of the illness in question. The free doctors, on the other hand, take no practical measures until they have learned about the disease from careful examination and consultation with the afflicted and his friends. The slave doctors never account for the causes or natures of the diseases they treat. The free doctors, by contrast, not only succeed practically, but they also discover explanatory accounts and pass along what they learn (Butler 2003: 331–332).
20. Mikhail Lomonosov was a Russian scientist and writer who made important contributions to literature, education, and science. He was the founder of the Moscow University (1755). Among his spheres of science were natural science, chemistry, physics, mineralogy, history, art, philology, optical devices, and others. Lomonosov was also a poet, and he created the basis of the modern Russian literary language. For more information please see Peter Hoffmann, *Michail Vasil'evič Lomonosov (1711–1765). Ein Enzyklopädist im Zeitalter der Aufklärung* (Frankfurt am Main: Peter Lang, 2011). Boris Menshutkin. *Russia's Lomonosov; Chemist, Courtier, Physicist, Poet*. Trans. from the Russian by Jeanette Eyre Thal and Edward J. Webster under the direction of W. Chapin Huntington. (Princeton: Princeton University Press, 1952).
21. "Sudebnik Ivana III" (Juridical Codex of the Tsar Ivan III) in *Rossiiskoe zakonodatel'stvo X–XX vekov* (Russian Legislation in 10th–20th centuries) in 9 volumes. Vol. 2. *Zakonodatel'stvo perioda obrazovanija i ukreplenija russkogo centralisovannogo gosudarstva* (Legislation of the Period of Formation and Consolidation of the Russian Centralized State), (Moscow: Juridicheskaia literatura, 1985), 56.
22. "Spisok s gramot tsarja Ivana IV k shvedskomu korolju" (Messages of the Tsar Ivan IV to the King of Sweden), in "Russkaja istoricheskaja biblioteka" (Russian Historical Library). Vol. XXII, (St. Petersburg: Juridicheskaia literatura, 1908), 42.
23. "Ot tsarja I velikogo knasja Mikhaila Fedorovitcha vsea Rusi v Sibir', na Verhotur'e voevode nashemu Maksimu Streshnevu" [From tsar and grand

prince Mikhail Fedorovitch to Siberia, Verhotur'e, to our voivode Maksim Streshnev] in *Akty istoricheskie, sobrannye i izdannye Arheograficheskoi komissiei* (Historical Acts published by the Comission on Study and Publication of Early texts). Vol. IV, (St. Petersburg, 1842), 403.

24. Pososhkov, Ivan. *O ratnom povedenii* (On Military Behavior), (Moscow, 1842), 280.
25. Vladimir I. Chuchmarev, *Leibniz G.W. i russkaia kultura* (Leibniz G.W. and Russian Culture), (Moscow, 1968), 19–20.
26. Yurii Kopelevitch, *Osnovanie Peterburgskoi Akademii Nauk* (Foundation of the Russian Academy of Sciences), (Leningrad, 1977), 87.
27. Smagina, 28.
28. Tatyana Moiseeva, "Fizicheskii cabinet Kunstkamery" (The Physical Cabinet of Kunstkamera) in *Priroda (Nature)*, 2003 (9): 95.
29. Moiseeva, 96.
30. Smagina, 28.
31. Feodor Politkovski (1756–1809)—a professor of natural history and practical medicine in Moscow University. Politkovskii, Feodor. "Slovo o svjasi istorii natural'noi s fizikoi, himiei i vrachebnym iskusstvom" (A Word on the Connection between Natural History, Physics, Chemistry and Medicine), 1796, P. 13.
32. "Otchet Shumakhera I. D. o pologenii del v Akademii" (A Report by Schumacher J. D. about the State of Affairs in the Academy in 1726–1727) in *Zapiski Imperatorskoi Akademii Nauk* (Memos of the Imperial Academy of Sciences), Vol. VII, (St. Petersburg, 1865), 15.
33. *Materialy dlja istorii Imperatorskoi Akademii Nauk* [Materials for the History of the Imperial Academy of Sciences]. Vol. I. (St. Petersburg, 1885), 35.
34. *Raspologenie uchenija the Emperor Peter II* [Disposition of Manoeuvres of his Majesty the Emperor Peter II], (St. Petersburg, 1728), 52.
35. This journal was published from 1728 up to 1742 in which various events taking place in Europe were reported on.
36. *Primechanija na Vedomosti* [Additions to Vedomosti], (St. Petersburg, 1733), 41.
37. *Experimental physics by Cristian Wollf which was shortened from German into Latin and then translated into Russian by Mikhail Lomonosov*. (St. Petersburg, 1746). The other factor which determined the fact that experiment was not simply borrowed in the Russian lexicon was the lack of the tradition of natural/experimental philosophy in Russia as it was the case in Europe in the timeof Francis Bacon and Robert Boyle.
38. Mikhail Lomonosov, *Sbornik trudov v poesii i proze* [The Collection of Works in Verse and Prose], Vol. I, 1757, P. 379.
39. "O instrumentah k poznaniju pogod" [On the instruments used to investigate the weather] in *Comments to the newspaper "Vedomosti"* (St. Petersburg, 1734). A source where information about scientific discoveries both in Russia and abroad were published.
40. *Geografija natural'naja. Nebesnyi i zemnovodnyi krugi* c *ih svoistvami I deistijami* [Natural Geography. Celestial and Amphibious Circles: Its Characteristics and Qualities]. Trans. from Latin by Fedor Polikarpov, 1736, P. 114.
41. *Kratkoe opisanie Kommentariev Akademii Nauk* [Brief Description of the Commentaries of the Academy of Sciences], 1728, Part I. P. 21.
42. Mikhail Lomonosov, "Slovo o polze khimii, govorennoe v publichnom sobranii Imperatorskoi Akademii nauk 6 sentabria 1751 goda" [Discourse on the Usefulness of Chemistry pronounced at the Public Meeting of the Academy of Sciences on September 6, 1751] in Lomonosov M.V. *Polnoye sobraniye sochineny AN SSSR* [Complete Works], Vol. 2: Trudy po fizike i khimii [Works on Physics and Chemistry 1747–1752], (Moscow, Leningrad, 1951), 354.

43. Mikhail Pankevitch, "Slovo ob otlicitel'nyh svoistvah i istochnikah prosvechenija" [A Word on the Distinctive Characteristics and Sources of Enlightenment], (Moscow, 1800), 47.
44. Ivan Shuvalov (1727–1798) was a favorite of the Russian empress Elizabeth, a patron of art and sciences.
45. "Pis'mo Mikhaila Lomonosova Ivanu Shuvalovu" [A letter of Mikhail Lomonosov to Ivan Shuvalov] in *Besedy v Obshestve ljubitelei rossiiskoi slovesnosti* [Discussions in the Society of Amateurs of Russian literature], Vol. 3, (Moscow, 1871), 76.
46. Mikhail Lomonosov, "Predstavlenie v Akademicheskoe sobranie o postroike khimicheskoi laboratorii" [A Report to the Academic Council on the Foundation of Chemical laboratory] in Lomonosov M.V. *Polnoye sobraniye sochineny* [Complete Works], Vol. 9, (Moscow, Leningrad, 1955), 52, Physical Chemistry°, ecii Naukит проблемы. Он, правда, написан на англ. языке. идеями и, поэтому, буду только рада, ес 23.
47. Mikhail Lomonosov, Prodromus ad veram chymiam physicam [Vvedenie v istinnuju fizicheskuju khimiju / An Introduction to True Physical Chemistry] in Lomonosov M.V. *Polnoye sobraniye sochineny* [Complete Works], Vol. 2, (Moscow, Leningrad, 1951): 505. Translation from Latin into Russian is done by a Russian scholar Boris Menshutkin (1874–1938), a chemist and a historian of chemistry, as well as a researcher who studied Lomosov's biography.
48. "there is a need for *isskusnyi* [skilled] *chemist* . . . the one who has learned science not basing only on books but who practiced it by his own iskusstvo" (meaning who has made a series of experiments). Lomonosov. Lomonosov, *Slovo o polze khimii*, 354. "v drugih Academijah kagdaja professija imeet v odnoi nauke dvuh ili treh iskussnyh (in other Academies every profession has three of four iskussnyh) (i.e. [skilled] experts)". Mikhail Lomonosov. "Vsenigaishee mnenie o ispravlenii Sankt-Peterburgskoi Akademii Nauk" [An Opinion on the Improvement of the Academy of Sciences] in Lomonosov M.V. *Polnoye sobraniye sochineny* [Complete Works], (Moscow, Leningrad, 1952), 22.
49. Lomonosov and other pioneers of natural sciences in seventeenth and eighteenth centuries, that studied artificially produced phenomena, could hardly foresee which results these experiments would lead to later and what fundamental concepts would be used together with the adjective «iskusstvennyi» like *artificial insemination, artificial Earth satellite, artificial intelligence etc.*
50. *Geografija natural'naja. Nebesnyi i zemnovodnyi krugi c ih svoistvami I deistijami* [Natural Geography. Celestial and Amphibious Circles: Its Characteristics and Qualities]. Trans. from Latin by Fedor Polikarpov. 1736, P. 197.
51. Geografija natural'naja, 170.
52. Georg Kraft, *Kratkoe rukovodstvo k matematicheskoi i natural'noi geografii* [A Brief Handbook on the Mathematical and Natural Geography]. Trans. by Ivan Golubtsov, 1739, P. 175.
53. The application of skill(s) to the arts of imitation and design, painting, engraving, sculpture, architecture; the cultivation of these in principles, practice, and results; the skilful production of the beautiful in visible forms.
54. Other Slavic languages, like Ukrainian, for example, shares a similar dualism—there are two terms for scientific experiment—дослiд (doslid) and експеримент (experiment) respectively.
55. The text of the original written by Lomonosov in Latin. "Experimenta alias instituta, quorum successus adhuc dubii sunt, cum debita attentione repetere". The English text is a translation from Russian, which in turn was done by Russian scholar Boris Menshutkin (1874–1938), a chemist and a historian

of chemistry, as well as a researcher who studied Lomonosov's biography. Lomonosov, Mikhail. "Predstavlenie v Akademicheskoe sobranie o postroike khimicheskoi laboratorii. October 25, 1745" [A Report to the Academic Council on the Foundation of Chemical laboratory] in Lomonosov M.V. *Polnoye sobraniye sochineny* [Complete Works], Vol. 9, (Moscow, Leningrad, 1955), 23.
56. Mikhail Lomonosov, "Proekt reglamenta Akademii Nauk, 1764–1765". [A Project on the Regulations of the Academy of Sciences] in Lomonosov M.V. *Polnoye sobraniye sochinenyi* [Complete Works], Vol. 10, (Moscow, Leningrad, 1952), 154.
57. Mikhail Lomonosov, "Programma" [Programme] in Lomonosov M.V. *Polnoye sobraniye sochineny* [Complete Works], Vol. 1, (Moscow, Leningrad, 1950), 535.
58. Mikhail Lomonosov, "Slovo o proishogdenii sveta, novuyu teoriju o tsvetah predstavlajuachaja, skazannaja v publichnom sobranii Imperatorskoi Akademii Nauk 1 ijulja 1756 goda" [A Word on the emergence of light, (ORIGINATING) presenting a new theory pronounced in the Public Meeting of the Academy of Sciences on July 1, 1756] in Lomonosov M.V. *Polnoye sobraniye sochinenyi* [Complete Works], (Moscow, Leningrad, 1952), 325.
59. The text of the original is written by Lomonosov in Latin. "Aquarum plures dari species, varietate corporum, quibus impraegnata sunt, distinctas, experientia docuit". In Russian the translation was done by Boris Menshutkin. Lomonosov Mikhail, "Vvedenie v istinnuju phisicheskyju himiju [An Introduction to True Physical Chemistry] in Lomonosov M.V. *Polnoye sobraniye sochineny* [Complete Works], Vol. 2, (Moscow, Leningrad, 1951), 519.
60. Mikhail Lomonosov, "Razmyshkenija o prichinah teploty i holoda [On the Causes of the Warmth and the Cold] in Lomonosov M.V. *Polnoye sobraniye sochineny* [Complete Works], Vol. 2, (Moscow, Leningrad, 1951), 648.
61. The text of the original is written by Lomonosov in Latin. "Hinc cavenda et expungenda sunt ea, quae in praeterito et praesenti statu Academiae bonis successibus noxia scientiarum per experientiam cognita habentur". In Russian the translation was done by Boris Menshutkin. Lomonosov Mikhail, "Predlozhenija ob organisacii i ustave Academii Nauk [Suggestions on the Organization and Charter of the Academy of Sciences] in Lomonosov M.V. *Polnoye sobraniye sochinenyi* [Complete Works], Vol. 10, (Moscow, Leningrad, 1952), 116.
62. Mikhail Lomonosov, "Predstavlenie v Kanceljariju Akademii Nauk ob otpravke nekotoryh studentov Universiteta za granitzu" [A Report to the Academy of Science about the Necessity to send a number of students to study abroad] in Lomonosov M.V. *Polnoye sobraniye sochinenyi* [Complete Works], Vol. 9, (Moscow, Leningrad, 1955), 595.
63. Mikhail Lomonosov, "Slovo o proishogdenii sveta, novuyu teoriju o tsvetah predstavlajuachaja, skazannaja v publichnom sobranii Imperatorskoi Akademii Nauk 1 ijulja 1756 goda [A Word on the emergence of light, presenting a new theory pronounced in the Public Meeting of the Academy of Sciences on July 1, 1756]. Lomonosov M.V. *Polnoye sobraniye sochineny* [Complete Works], Vol. 3, (Moscow, Leningrad, 1952), 320.
64. Mikhail Lomonosov, "Proekt reglamenta Akademii Nauk, 1764–1765" [A Project on the Regulations of the Academy of Sciences] in Lomonosov M.V. *Polnoye sobraniye sochinenyi* [Complete Works], (Moscow, Leningrad, 1952), 141.
65. Mikhail Lomonosov, "Vvedenie v istinnuju phisicheskyju himiju (An Introduction to True Physical Chemistry" in Lomonosov M.V. *Polnoye sobraniye sochineny* [Complete Works], (Moscow, Leningrad, 1951), 513.

66. "Opyt o vospitanii blagorodnyh devits". Perevod s franzusckogo jazyka na russkii Mikhailom Semchevskim [An Essay on the Education of the nobles young ladies. Translated from French by Mikhail Semchevskiy], (St. Petersburg, 1778). "Opyt o providenii bogiem". Perevod s franzusckogo jazyka na russkii svjashennikom Ioannom Sidorkovskim [An Essay on Providence. Translated from French by priest Ioann Sidorkovski], (St. Petersburg, 1781).
67. Reinhart Koselleck, *Futures Past: on the Semantics of Historical Time*. Transl. And with an Introduction by Keith Tribe (New York: Columbia University Press, 2004), 86.
68. Martin Jay, *Songs of Experience. Modern American and European Variations on a Universal Theme* (Berkeley and Los Angeles: University of California Press, 2005), 12.

9 Linguistic Precision and Scientific Accuracy

Searching for the Proper Name of "Oxygen" in French, Danish, and Polish

Jan Surman

1. INTRODUCTION

The relation between language and science has been a matter of inquiry since the very beginning of scientific activity.[1] Since Bacon's critique of words as idols of the marketplace, the issue of proper scientific terminology has been linked with the idea of scientificity as a whole, from ideas of artificial languages to the quest to improve vernaculars. Historically, this discussion intensified at the time vernacular languages began to challenge Latin as the scholarly tongue in the early modern period. The de-Latinization movement of the seventeenth century was seen as abandoning the scholastic manner of understanding the world, linking secular scientific ideas with nationalist ideologies; in the case of nationalistically based linguistic reforms, foreign words were to be expunged, for example traces of French in German.[2] This so-called linguistic purism has not yet found much place in the discussion of how the sciences have developed, although at the time it was an issue scientists were confronted with and which they discussed. They did so both within their own language community but also beyond it, such as in the first international meeting of chemists (the Karlsruhe Congress) held in Germany in 1860.

Chemistry, with which this chapter is principally concerned, has close ties to vernacular languages for its own historical reasons: first, chemical substances were being discovered and named for years before the advent of modern chemistry in the second half of the eighteenth century, leaving distinct residues of these names in every language; second, chemistry largely outgrew a "professional" discourse linked to industry that during the Industrial Revolution became widespread. This discourse was both aloof of academic discourse yet intertwined with it.[3] Moreover, in the late eighteenth to early nineteenth century, an international community of chemists was already writing in vernaculars, often using words denoting the properties of the substances/elements they discovered and stabilizing their experimental systems through appropriate nomenclature. For example, Joseph Priestley (dephlogisticated air), Carl Wilhelm Scheele (Feuerstoff), and Antoine Lavoisier (oxygen) named the same substance in three different ways, each adapted to the theoretical framework they used: respectively, phlogiston theory, *Feuerlehre* (theory of combustion), and the theory of acidity. By 1800 the French chemists' nomenclature was most widespread and the names they used were standard in chemistry, having been translated, appropriated, or simply used without change by the scientific community.

In what follows I trace one of the terms Lavoisier used—*oxygène*—from its first proposal in the late eighteenth century until the 1850s, when alternative names were adopted in Danish and Polish. As will be explained, the naming and re-naming of the element can only come from a clash between chemical theory and applied concepts of scientific language facilitated by both the inner-scientific politics and general cultural trends of the époque. The extensive discussion of chemists rationalizing their choice gives insight into how these scholars perceived language. Many of their ideas and notions are still present in the current terms themselves and show that vocabulary we now consider abstract was seen as much less so in the nineteenth century. The reification of vocabulary that I examine in this chapter was not only a tinkering with language: it aimed at the betterment of the contents of knowledge through changes to its medium of representation.

2. NEW VOCABULARY FOR NEW CHEMISTRY: FRENCH CHEMICAL GRAMMAR

As far as "modern" chemical terminology is concerned, there is probably no controversy about its origin. The terminology can be traced back to Paris between 1770 and 1790, with Antoine Lavoisier as the pivotal (albeit not the sole) figure responsible for its formation. The Paris of those years was a fuzzy entre-regime capital where various approaches to science coexisted[4] and where this in-between state of affairs nurtured scientific language reforms in terms of vocabulary. In several disciplines, these oscillated between rendering vocabulary in the vernacular and using classical languages as the base for new nomenclatures.[5] This revolutionary atmosphere affected chemistry too. In this field new nomenclature was devised concurrently with new experimental and theoretical discoveries. One of the most durable scientific results of this period was the rejection of phlogiston theory, for which Lavoisier is mostly credited. As a result of this, the nomenclature of chemistry had to be substantially reformed.[6]

The Method of Chemical Nomenclature (Méthode de nomenclature chimique), published in 1787, was a joint publication by Lavoisier, Louis-Bernard Guyton-Morveau, Claude Bertholet, and Antoine François Fourcroy which established the linguistic architecture of post-phlogiston chemistry. Its key point was the systematization of disciplinary vocabulary in order to create a stable design through which future discoveries could be seamlessly integrated.[7] The authors proposed using Greek terminology as the basis of new terms. Fourcroy when writing on this issue stated that he wanted to reduce the amount of Swedish and German terms, as well as those of scholarly Latin, the former because they were incompatible with French, the latter to symbolize the emancipation from churchly influences. After the revolution, he added that aesthetics played a role too: Greek words sounded pleasant and were easy to pronounce, coming near to the "genius of the French language".[8] While the "nomenclaturists", as they were later called, did not claim to be devising an artificial system but rather to be revising, enlarging, and systemizing the existing vocabulary by removing confusing terms or various terms denoting one substance, the linguistic reality was different. The names of the elements which eventually gave basic structure to the arrangement were neologisms.[9] Translations of the *Méthode de nomenclature chimique* were rapidly made: it was rendered into English and Spanish in the next year, into Italian in 1790, and into German in 1793.[10] The system of classification used in the book was also included in the next edition of the Encyclopedia Britannica (1791–1799), which presented it as accepted knowledge.[11] The rapid spread of this work indicated that the need for a structured vocabulary was not exclusive to France. While not instantly and uncritically accepted, this approach to chemical name-giving was facilitated in particular by Lavoisier's own achievements and publications.

Lavoisier, with whom the reform of the chemical nomenclature is commonly associated, propagated the new system in his *Traité élémentaire de chimie, préséinte dans un ordre nouveau et d'après les découvertes modernes* (Elementary Treatise on Chemistry, Presented in a New Systematic Order and Alongside Modern Discoveries)[12] (1789), which, as the first large overview of the "new" chemistry, went through a wave of translations that served to stabilize the new vocabulary.[13] Famously, the introduction to the book was a hymn of praise for the French grammarian Etienne Condillac and his approach to language, which Lavoisier eagerly adopted.[14] Condillac stood in the tradition of the grammar school which advocated the possibility of perfecting and stabilizing the language according to philosophical ideals. Jean-Baptiste le Rond d'Alembert in *Éloge de M. du Marsais* (Encomium of M. du Marsais) was one of the first thinkers to express this philosophy. He opened the seventh volume of the famous *Encyclopédie* stating that "it is [a duty] of philosophers to regulate the language as it is for fine writers to stabilize it".[15] The regulation and disciplining of natural language was, in contrast to artificial languages, based on the vernacular and popular speech. However, as a Lockean sensualist claiming a direct relation between an object and its denotation,[16] Condillac did not go so far as the German philosophers in claiming the reformation of speech on the basis of *Volkssprache*. Yet he did criticize the distinctively conservative language of elites, which in his eyes simply represented political, social, and religious supremacies that were blocking a reform process.[17] In his textbook *La logique ou les premiers dévelopments de l'art de penser* (Logic or The Early Development of the Art of Thinking) (1780), Condillac elevated language as a tool in scientific reasoning. Language allowed one to methodically arrange and structure thoughts. His notions of the efficacy of language as a tool took its cues from ideas of scientific exactitude—algebra was for Condillac the ideal of well-ordered language, and scientific language(s) that were based on his ideas were consciously modelled as such.[18] The metaphor of language as an instrument was also used by Lavoisier in his contribution to the *Méthode*: "If languages are truly instruments formed by the people to facilitate the operations of their mind, it is important that these instruments are as good as possible".[19]

Lavoisier not only remained close to Condillac's ideas on the language-science relation, but also enlarged the scope of the desired influence of language to incorporate epistemic qualities. A systematic, well-ordered language would allow one to sense the not-yet-discovered substances, thus serving as a kind of material instrument.[20] According to Lavoisier, nomenclature could not be separated from its scientific discipline since words are direct connectors to ideas—just as few years earlier Condillac had taken science and language to be inextricably connected, to the extent that the reform of one necessarily produced a betterment of the other. Yet Lavoisier envisaged a non-artificial nomenclature, and his claim was to retain the old names as long as they were appropriate or to adjust them into the new logical system. The latter implied the need for manifold innovations, mostly through the use of Greek, a language highly esteemed in the sciences of pre-revolutionary France.

Among the newly devised terms, the authors proposed new names for the main elements—*oxygène*, *hydrogène*, and *azote*. The names were selected on the basis that they denoted the specifics of these elements. Oxygen was a substance which Lavoisier deemed responsible for acidity; thus, the name was derived from Greek ὀξύς (acid) and γείνομαι (to engender)—from these two stem *oxygenation* and *oxides*. Hydrogen was a compound of water and was thus formed with the stem of *ὕδωρ*. In the naming of nitrogen, however, Lavoisier was cautious not to ascribe any general chemical properties, as the gas's properties were "hitherto but little known".[21] Consequently, the name chosen, *azote* ("from the Greek privitive particle α and *ζωή*, vita [life]"),[22] expressed the property of depriving animals that breathe it from life. Lavoisier considered another possible name, *nitrigen*, as the element that formed nitric acid, but as it was not confirmed on a general basis, he opted for *azote*. The name was, however, not universally accepted, and eventually the French chemist Jean-Antoine Chaptal successfully popularized *nitrigenè/nitrogene* as forming the nitre (saltpeter). Nonetheless, the names *azote*, *nitrogen*, and *alkaligen* were used simultaneously for some years—with contemporary French and some other languages retaining *azote* and others adopting *nitrogen*.[23]

The adoption of this new chemical vocabulary varied from language to language. In English, for instance, the three compounds in question were adopted, with *azote* being later changed to *nitrogen*, although the adherents of phlogiston firmly kept their terminology and rejected French innovations.[24] Similarly, in Spanish (in which language the word *azote* meant whip) the words *azoote/azoe/ázoe* grew common until the second half of the century, when *nitrogen* prevailed.[25] German scholars adopted French terms as well, but translated them into *Sauerstoff*, *Wasserstoff*, *Stickstoff*, changing the active particle-gene into *Stoff* (= substance, matter), yet retaining the term *oxidation* (as *Oxidierung*), and thus in a way rupturing the linguistic unity of the French terms. These terms were used in the review of *Méthode* by (probably) Sigismund Hermbstädt, who translated the indivisible substances (the first class mentioned in the book) as lumière = *Lichtstoff*, calorique = *Wärmestoff*, oxygène = *Säurestoff*, and hydrogène = *Wasserstoff*. The fact that he did so shows these were perceived as elements. For *azote*, Hermbstädt used *Tödlicherstoff* (from *Tod* = death).[26] German translators used thus the loan rendition to retain the characteristics Lavoisier pledged, but also to adapt them to the then developing German scientific language, which at the time was liberating itself from French influences. In the next years, especially the *Tödlicherstoff* was replaced by the nativized form *Stickgas/Stickstoff* (*ersticken* = to suffocate, to deaden) as more convenient linguistically. Hermbstädt, the translator of Lavoisier, preferred, however, *Salpeterstoff* or *salpetererzeugender Stoff*, a clumsy translation of the French *nitrogène*. *Azote* (and thus all its linguistic derivatives) points to characteristics which better well-known gases possessed and because of this lacked precision.[27] *Stickgas* was here used by Chantal's translator, Friedrich Wolff, who noticed that, although the name was not wholly correct, neither was *nitrogène/Salpeterstoff*. It was not saltpeter but nitric acid that was created by nitrogen. In addition, *Stickgas* was shorter than *salpetererzeugender Stoff*.[28] Thus, the change of *azote* to *nitrogen*, which was facilitated by Chaptal through the translation of his work into English, was further inhibited in German because in the German version of Chantal's treatise the reasons for his selection were rendered invisible by the translator's choice. Wolff rendered *nitrogène* as *Stickgas* and thus over-wrote Chaptal's linguistic-chemical innovation.

The German approach to the translation of these terms was not unique. The first authors writing in Czech in the 1820s (Friedrich/Bedřich Wšemír von Berchtold and Jan Swatopluk Presl) used a similar approach to loan translations in their books on botanics and employed *kyslík* (*kyselina* = acid, *-ík* = active participle, he who gives/makes), *vodík* (*voda* = water), and *dusík* (*dusit* = suffocate, strangle).[29] Presl used the same words in the first textbook on chemistry in Czech, which he composed in 1828.[30] The community of chemists actively using the Czech language was small up until the second half of the century—consisting primarily of Presl and his students—resulting in fewer controversies than in other languages. The latter reforms linked with Karel Slawoj Amerling and Vojtech Šafárik did not touch these three elements. Other authors adopted the terminology, which was also codified in Josef Jungmann's authoritative German-Czech dictionary of 1833.[31] Some modifications were proposed, for example by the physician and natural scientist

Filip Stanislav Kodym, who, using Presl's terminology, asserted that, because of their properties, oxygen could also be called *živík* ("he who gives life") or *hořík* ("he who gives fire") and nitrogen, *mírník* ("he who gives calmness"). In the latter case the reasoning was that breathing pure oxygen gas would cause vigorous though potentially deadly activity, and only through nitrogen could this effect be calmed.[32] The suffix *-ík* ("he who gives") also implies that the "base" language for this terminology could not have been German. Indeed, Presl referred to the Polish terminology of Jan Śniadecki as his source.[33] While the politics of de-Germanization is one factor leading to this choice, another might be Pan-Slavic thought and the possibility of creating a Slavic literary language with dialects (such as Czech or Polish). This was not an unpopular idea at the time. For instance, the first Czech journal, *Krok*, edited precisely by Presl, was subtitled Czecho-Slav.[34] The distinction between acids and oxides was rendered into the vernacular through acid (*kyselina*) and oxide (*kysličnik*), so that *oxidation* became *okysličení* and *reduction*, *odkysličení*.

The remarks on the terminological meanderings of these three elements demonstrate that the process of reforming language was approached differently by various scholars. While the idea of a new language for the new science was accepted, the formalities of language were matters of mediation between Greek-French terms and their justification, between the existing terminology and associations readers would make, and finally between the characteristic qualities of the described substances. The solutions differed between languages and individual authors and changed as well with the advent of more exact descriptions of element functions. The convention applied by the French chemists of distilling the name from its best known properties, perpetuated particularly in the case of nitrogen, produced differing results: a vernacularized German expression on one side versus the simple borrowing of words in English on the other. For most of the nineteenth century, more than one term was used, which also shows that the way of translating the elements was not necessarily a single-person decision, but had to be validated in practice. Some points of criticism coincide in different languages; for example, the use of *azote* instead of *nitrogen* was considered to disrupt the continuity of nomenclature by identifying a gas by a particular property that many others shared.[35]

3. THE CHEMICAL TRUTH IN LANGUAGE: POLISH AND DANISH

Even if the three terms (*oxygène*, *hydrogène*, *azote*) were partially adopted, this adoption was not universal. Throughout the century, the linguistic system devised in France was prevalent, although with two chief exceptions. In Danish and Polish the discussion over terminology generated a reconsideration of the most vital compound in French taxonomy, *oxygène*, and led to its replacement with another term. In both languages I will analyze more thoroughly the prevailing logic of linguistic purism that guided these decisions. Both the process of term change and its justification in both these cases show many similarities.

Hans Christian Ørsted (1777–1851), who was responsible for the rejection of *oxygène* in Danish, is mostly known due to his experiments with electromagnetism. However, his interests were very broad and, especially in his younger years, were not confined to the natural sciences. He was a well-educated romanticist with a strong interest in languages and literature (himself being a poet, translator, and writer) and a profound organizer of scientific infrastructure during Denmark's so-called Golden Age (*danske guldalder*), that is, the period of artistic and scientific flourishing of the first half of the nineteenth century.[36] His particular Kant-informed view of scientific language can easily be called "moderately puristic",[37] although with one reservation. While purism normally denotes concentration on one ("national") language and claims of its cultural and historical independence, Ørsted called for a Nordic language with dialects (a model taken from German).[38] His particular attention to Norwegian suggests this pan-Scandinavism also had a strong political dimension, as Denmark lost Norway to Sweden several years earlier. The claim of Norwegian-Danish linguistic affinities had imperialistic connotations.

In 1814, already as a professor in Copenhagen, Ørsted published *Tentamen nomenclaturae chemicae omnibus linguis scandinavico-germanicis communis* (Attempt to Chemical Vocabulary Common for All Scandinavian-Germanic Languages), which outlined his approach for the reformulation of existing Greek-French terminology and its replacement with Germanic names.[39] The same year the brochure was translated into German and published in the renowned *Journal für Chemie und Physik*, but it was scarcely noticed, and the proposed ideas left almost no durable traces. New names for oxygen and hydrogen were, however, accepted in Danish and remain at least partially in use up to now. This was not the first attempt to propose new nomenclature, which Ørsted already used in his unpublished manuscript *Kraftlæren* (Theory of Force), but for some reason it did not reach the printing press until recently.[40] In the manuscript he promoted a vocabulary closely tied to combustion theory—influenced by Johann Wilhelm Ritter and József Jakab/Joseph Jacob Winterl's chemistry. He combined a romantic approach to *Naturphilosophie* with modern developments (e.g., Winterl's opposition between acidity and alkalinity). Thus, as in Lavoisier's case, new nomenclature was linked with an anticipated revolution in chemistry.

In the *Tentamen nomenclaturae* Ørsted concentrated, however, on another pivotal issue of the terminology, scarcely mentioning his new ideas on chemistry. His point of concern was that through the use of compound words the chemical nomenclature in Germanic and Scandinavian languages did not distinguish between elements and (chemical) compounds. More especially they did not easily allow the formation of derivatives to describe chemical processes and classes of compounds when compared to the (in his eyes) nonetheless clumsy classifications based on *Sauerstoff* (e.g., *Halbsäure* [lit. half-acid] as oxide).[41] Thus, instead of using native terms, chemists writing in German used words like "*oxidieren, oxidiert, Oxidation, desoxidieren*".[42] While Ørsted understood that this would facilitate communication across languages, it hindered understandability by non-experts, those "for whom he writes".[43] Concerning this point, he also used emotional metaphors to underscore both the value of language for each nation and his concept of Germanic affiliation, which was based on the notion of "purity of language, which every nation (*Volk*) has to guide like a Palladium",[44] and the existence of "brother-languages" that reflect "a holy tie of blood relationship" (referring to Danish and German).[45] His six points defining good nomenclature were thus a mixture of philosophical and linguistically purist arguments:

> 1) the names have to be *true*, i.e must not fix false concepts of things . . .; 2) the name has to *denote* the thing and not *describe* it, if it cannot be done through display of its [the thing's] elements; 3) where possible, especially while naming the elements from which many substances can be built, one has to beware that the names of the elements are not compositions (like e.g. in words *Wasserstoff, Vandstof, Selpeterstoff*) . . .; 4) . . . names have to be *fruitful*, i.e. so many names can be deduced from one, as from the named thing other [things] can be produced; 5) In the building of excellent names, one has to pay attention that the *composites* are not too long or hard to pronounce; 6) Foreign words are to be rejected, *if they cannot be adapted to the grammatical rules of the mother tongue*.[46]

Other criticism was directed against the term used for oxygen, or rather, against its "Germanic" translation—*Sauerstoff, surstof*—which in Ørsted's eyes covered only one of the properties, leaving the alkalides and oxygen-based chemicals aside. Elaborating on the properties of the element, Ørsted called *sauer* and *scharf* (sharp/biting) its most important characteristic. Only the first one was used in translations into Germanic languages. The Greek word, on the other hand, embraced ("by chance", as he stressed) both properties.[47] Moreover, most of the aforementioned rules for designing good terms were violated (especially 2 and 3), which prompted Ørsted to search for a new term of Germanic origin. In accordance with combustion theory, the main property of oxygen was not its acid-building qualities, but its role in burning. He rejected *Feuern* (from fire), however, as linguistically inappropriate for building both a term and its derivatives like oxygenation. He thus proposed an adopted Danish word for fire—*Ild* (Ilt), which in various forms existed in other Germanic languages (e.g., Icelandic *Eldr*, Swedish *Eld*, Anglo-Saxon *Älled*) and is found in the "Mother of our [languages]", Persian.[48] Taking a detour through Greek he also linked *ild* with the Swedish word *ala*, which also meant "to create, to produce . . . to support",[49] thus highlighting "a trace of concealed truth, which one can find in the oldest names of things".[50]

The changes proposed by Ørsted went of course beyond reforming the name of this particular element. With similar argumentation, hydrogen was changed into *brint*, derived from burning. These alterations aimed to simplify the whole chemical nomenclature by deriving familiar-sounding and more understandable names for compounds. It also aimed to do away with the previously mentioned incongruity of German *Sauerstoff-Oxidation*, which would now become *Elt-Eltung* (*Ilt-Iltet* in Danish), and *Halbsäure* (Oxide), which would be *Geelt* (G) or *Ilte* (D) when non-oxidized and *Geeltet* (G) / *Iltet* (D) when oxidized. Ørsted, through his word choice, thus linked the need for a linguistically structured system with the purist argument for the value of national tongues—which since the Romantics, and especially since Herder, had grown in importance in philosophy—to his ideas of combustion. Notwithstanding the lack of direct reference in this particular text, combustion was still central to his chemistry.

The aforementioned changes in the vocabulary proposed by the Danish scholar did not influence the nomenclature in Scandinavia and Germany, although *ilt* and *brint* are still used in Danish and partially in Icelandic.[51] They, however, resonated with Polish scholars. In 1853 Jan Oczapowski (1800–1871) proposed a change in Polish nomenclature by indicating that the Danish Learned Society had recently adopted similar changes as binding. Adoption of new terms for oxygen in Polish does show, however, slightly different traits from the adoption in Danish.

Discussions over the shortcomings of the application of French terminology to Polish were particularly long and complex, and the question of oxygen and other element names was only part of the problem. From the early nineteenth century until the early twentieth century, scholars and chemists repeatedly discussed this issue, finally creating in the 1870s an interdisciplinary commission at the Cracow Academy of Sciences and Arts to create a binding nomenclature by the end of the century. The reason for this long span of time was lack of authoritative institutions and scarcity of contacts among the Polish-speaking chemist community, which was scattered through three empires and only in the later nineteenth century had begun to enjoy state support in Habsburg Galicia. Only then did Polish become a language of instruction in higher education. This created a necessity for textbooks to transmit knowledge in a generally agreed terminology. The larger context of the dispute was the relation between language regulation and vernacularization in questions of compound systematization, that is, affixation as well as denominations of the elements which were the base for the more complex designations. From the beginning to the end of the century, several systems were devised to meet both scientific and linguistic standards, quite often based on adaptations of French, German, or Latin systems. Two differing approaches emerged in Warsaw and Cracow in the second half of the century.[52]

In contrast to other issues, the change of designation of oxygen was easily agreed on. Discussions took place in the 1830s and 1840s and symbolically concluded in 1853 (although several names continued to be used simultaneously afterwards). As in the Danish case, the argumentation moved from an initially used loan translation to a vernacular word which was thought would refer more directly to oxygen's functions and properties. The process was much more in line with the general trend towards linguistic engineering than with the singular arguments of Ørsted. The first name used in Polish was not as in the Danish case modulated after German, but directly after French. It was proposed by Jędrzej Śniadecki (1768–1838), professor at the Polish-language Vilnius University in the Russian Empire.[53] In alignment with the French tradition, he suggested *kwasoród* (*kwas* = acid, *ród* [from *rodzić*] = one who creates) for oxygen, *wodoród* (*woda* = water) for hydrogen, or *saletroród* (*saletra* = saltpeter). For nitrogen the name *dusik* (one who suffocates, strangles) was used in correspondence with *azote*.[54] As far as semantics was concerned, Śniadecki was a conventionalist about the word-language relation and linked this idea with purism and vernacularism by claiming that words used in the common speech should be used and appropriated if necessary only if such constructions enabled scientific work and communication. According to him it was much easier for students (but also for scholars) to imagine an element if they could read its properties from its denomination—although due to the transfer problem he used Latin equivalents in brackets in his text. The aforementioned question of classification became the first point of conflict between Śniadecki and Warsaw chemist Aleksander Chodkiewicz, who criticized the Vilnius scholar for neglecting the classificatory function of the chemical nomenclature with inconsistent use of endings for acid/weak-acid while there were more apt Polish words.[55] In his response, Śniadecki stressed that it was of greater importance that the names of compounds related to commonly used language than to abstract rules of classification (which he nevertheless adopted by using the differentiation *kwas/podkwas* to denote acid/weak-acid instead of relying on ending only). This interesting exchange of ideas was essentially a debate over the "best" Polish words—"best" in the sense of providing more information about the elements themselves—in relation to the newer theoretical developments in chemistry which required adjustments to the terminology of the first publications on new chemistry in the previous years.[56]

The publication which changed the Polish denomination of *oxygène* was not directly part of the purist movement, although it shared several of its features. Vilnius graduate and Warsaw physician Jan Oczapowski, whose 1853 paper on the name of oxygen was part of one of the first commissions that aimed at rewriting Śniadecki's vocabulary, referred directly to his famous predecessor and his teacher in Vilnius when proposing his amendment.[57] In the introduction to his article he extensively quoted the Vilnius chemist's ideas on scientific language and its practical embedding (within industry, agriculture, and medicine) and thus the necessity to keep the nomenclature near to commonly used language. While Oczapowski shared Śniadecki's views and quite openly revered him, he described the change of nomenclature as of utmost necessity. The question of oxygen was in his eyes especially pivotal: "To free us as soon as possible from the labyrinth which *kwasoród* led so far, to light up the temple of such an excellent science and to keep the light of truth, one necessarily has to begin not only to remove *kwasoród*, but also all other words ending with *ród*".[58] The reason for it—most vivid in the case of oxygen but also in that of other terms—was the imprecision of the term. Several substances had been identified as responsible for creating acids, and thus the exclusivity of oxygen as acid-bearer, which was implied by *kwasoród*, was simply erroneous and would lead to confusions in education and in praxis. Instead of the translation of the French term, he proposed a better consideration of the properties of the element when naming: "i.e. that with the help of oxygen all substances begin to glow, burn and turn into dust, because all what is combustible joins with oxygen [. . . and thus] glowing and turning sour are two words marking the properties of one substance, *oxygenium*".[59]

Oczapowski thus proposed a derivative from glowing (*tlić*)—*tlen*. In order to claim the legitimacy that a connection to such a prominent chemist could provide, he "tested" his proposal against the rules of Śniadecki. Most importantly he stressed that the word had not existed hitherto and that it was not only Polish but could be found in other Slavic languages, with a meaning in Old Slavic "to turn into dust, to decay, to rot etc."[60] Thus, without directly referring to the Danish case—which he certainly knew, noting in a footnote at least the change of *surstoff* to *ilt*[61]—Oczapowski followed a similar argumentation as well as the idea of "concealed truth" which Ørsted had sought in languages of the past. Similarly, as in the Danish case, the ease of using the new name in classifications and descriptions of compounds and processes was of importance. The removal of the association between acids in oxides (*tlenek*, previously *niedokwas*) and oxidation (*tlenienie*, previously *kwasorodnienie*) was especially underlined. This served to transform the idea of acids themselves, which were divided into *kwasy* (acids) and *tlenokwasy* (acids with oxygen), thus embedding recent chemical discoveries.

The major difference to the Ørsted, however, is that Oczapowski was not interested in revolutionizing chemistry, but primarily in adapting Polish language to the new knowledge. It might be easy to place him under the banner of purism, especially with his references to "right" and "false" terms and stressing the value of the vernacular, nationalized names in education. However, there is little to support this philosophical alignment, and the change of terms is from Polish into Polish, from Greek-French loanwords in Śniadecki, to Old Slavic in Oczapowski. It would also be misleading to simply assume that Oczapowski was choosing a term different from the Russian one (*kyslorod*) only for political reasons: he claimed that his change would spread than into other Slavic languages. Yet the acceptance of the term may have been driven by anti-Russian sentiments, which grew considerably after 1863.

It took a few decades until *tlen* was officially adopted. Tradition opposed this novelty, but also the fact that the previously used *kwasoród* was closer to French and hence facilitated better trans-linguistic communication.[62] Even scholars who disagreed with *tlen*, like one of Warsaw's leading industrialists, Karol Lilpop, preferred it notwithstanding the shortcoming because it was more flexible in creating derivatives—even if the linguistic-conceptual idea was not precise.[63] Emilian Czyrniański, at this time a professor of chemistry in Cracow, criticised *tlen* for not being easily linked with the concept of acid. It would require changing the system of chemical compounds that he had devised. In his eyes, the fact that *kwasoród* gave false impressions about the element was widely known: the language of chemistry was abstract, and thus the connection between the word and its properties was not of much importance. Yet, despite this, he also changed his system in the 1870s.[64]

4. CONCLUSION

From the first naming of chemical elements, the discussion around them took an interesting turn which illuminates much about the understanding of science at the time. Chemistry was clearly regarded as an intersection between the experimental and linguistic-descriptive components of science, which could not be easily separated. Words not only denoted the object and its properties, but in a structured system of chemistry had to enable future discoveries, in the same way as mathematical systems of signs were thought to do. But to be capable of doing so, they had to reflect the object they denoted correctly and precisely by concentrating on its "unique" properties. This idea of uniqueness, however, conflicted with experimental change and scientific progress and the need for changes of nomenclature every now and then when new developments arise. At the same time, tradition and actual use of vocabulary, its codification in material media like books, makes nomenclature resistant to change. Thus Lavoisier's nomenclature had been accepted and unchanged even though criticized due to its inaccuracies by virtue of his experimental successes. The discussion on chemical terminology continued, although it became increasingly more concerned, from the 1860s and the Karlsruhe Congress up to the establishment of the International Union of Pure and Applied Chemistry after World War I, with the question of compounds. This discussion can be linked with the changing the role of classical languages in education: terms which French chemists of the eighteenth century saw as representations of properties became associated with the elements because increasingly fewer scholars associated the roots of the terms with Greek words. Treatment of the names of the elements as abstractions prevailed in languages which used loan translations. Thus, words with imprecise meanings, like the German *Sauerstoff*, remained in use.

While one could argue there is no such thing as a false term, *oxygène* and *azote* do not correspond well to the basic assumptions made of them:

oxygen is not a necessary compound of acids, and azote is one of many deadly gases. One could say that the creators of French chemical terminology would have used different names several decades later had they known the properties more accurately. In fact one of the reasons why the Greek-French terminology was kept in so many languages was because its associations were not so well understood and therefore almost had a life of their own. That the Greek terms led to some inaccuracies in the new chemistry is clearly perceivable if one considers the words *oxides* and *acids*, which describe two different classes of substances with the same designation but in two different languages. This distinction had already been highlighted by Lavoisier. Looking closely at the chemistry of that time, some names already seem too imprecise to stand for the elements. One can think of the German term *Feuerstoff*, which, depending on the writer and his theoretical approach, meant, among others, oxygen, hydrogen, or caloric, since the property-led nomenclature could be related to any of these elements. Such deliberations were carried out quite frequently.[65]

On the other hand, the developments concerning oxygen in Danish and Polish illustrate opposing ideas of abstract versus representational uses of scientific terms. In contrast to French or English, in the languages we have discussed the idea of representation prevailed. Terms were structured according to the notion of common linguistic knowledge and the philosophically led conviction that terms had to reflect reality in a way understandable to the broad public in order not to hinder scientific understanding and the progress of the nations. The question of terminological change also gave these scientists opportunity to rethink the place of oxygen in the chemical system anew and incorporate new developments in its naming. The relation between chemical terminology and chemistry was thus dialectical. While there is a trace of Romantic exaggeration in the form of the idea that truth is hidden in past language, for both Ørsted and Oczapowski this idea provided more of an auxiliary argument rather than a point of departure. They did not intend to uncover the "truth" about the element from past books, but rather express the experimentally acquired "truth" of the element with the help of the past. Ørsted even called his language a *Kunstsprache*—constructed language—based on present and ancient vocabulary.

To a certain extent, one could label the strategy of Lavoisier, as well the strategies of Ørsted and Oczapowski, as "epistemic purism". While linguistic purism tries to replace words labelled as foreign by vernacular, epistemic purism aims to perfect knowledge and the processes of its future production by replacing all terms associated with the previous "knowledge regime". Lavoisier, Morveau, Bertholet, and Fourcroy were clearly more successful when putting forward a terminology that removed all the terms associated with phlogiston. Both Ørsted and Oczapowski essentially followed the same approach to knowledge betterment through management of language.

However, Lavoisier and scholars adopting his terminology regarding foreign languages did differ in one respect. French scholars used Greek terms to name pivotal substances, rendering them abstract. But at the same time, as Fourcroy stated, they considered them similar to French. Ørsted, Oczapowski, and scholars in languages like German or Czech chose terms which were understandable for speakers of the vernacular. This marks a change between the still classicist sciences of late eighteenth-century France and Central European approaches. The latter were at least partly influenced by Herder's linguistic theories and aimed at a wider public which would require more easily understandable language.[66]

NOTES

1. The draft version of this paper was discussed at the *Wissenschaftshistorisches Kolloqium* at the Goethe-University in Frankfurt on November 6th 2012. I want to thank all participants at this colloquium for their insights. For the comments to an earlier version of this paper I thank Michael Gordin, Miles MacLeod, Rocío G. Sumillera, and Ekaterina Smirnova.
2. Wolfgang Walter Menzel, *Vernakuläre Wissenschaft. Christian Wolffs Bedeutung für die Herausbildung und Durchsetzung des Deutschen als Wissenschaftssprache* (Tübingen: Niemeyer, 1996).
3. Cf. Maurice P. Crosland, *Historical Studies in the Language of Chemistry* (Mineola: Dover Publications, 2004); Anita Kildebæk Nielsen and Soňa Štrbáňová, eds., *Creating Networks in Chemistry: The Founding and Early History of Chemical Societies in Europe* (Cambridge: The Royal Society of Chemistry, 2008).
4. See most recently Mi Gyung Kim, *Affinity, That Elusive Dream: A Genealogy of the Chemical Revolution* (Cambridge: MIT Press, 2003).
5. On the esteem of Greek in France see e.g. Peter Pickard, ed., *The French Language in the Seventeenth Century: Contemporary Opinion in France* (Cambridge: D. S. Brewer, 1992); Pietro Corsi, "After the Revolution: Scientific Language and French Politics, 1795–1802", in *The Practice of Reform in Health, Medicine, and Science, 1500–2000*, eds. Margaret Pelling and Scott Mandelbrote (Aldershot: Ashgate, 2005), 223–45; John Considine, "Ancient Greek Among the Eighteenth Century Languages of Science: Linnaeus, Dillenius, and the Lexicographical Record," *International Journal of the Classical Tradition* 16, no. 3–4 (2009): 330–43.
6. For a concise historical overview of chemical language revolution, its criticism, and effects see Marco Beretta, *The Enlightenment of Matter: The Definition of Chemistry from Agricola to Lavoisier* (Canton, Mass.: Science History Publications, 1993), 159–322.
7. Antoine Lavoisier, Louis-Bernard Guyton-Morveau, Claude Bertholet and Antoine François Fourcroy, *Méthode de nomenclature chimique* (Paris: Cuchet, 1787). More thorough analysis of the language of chemistry can be found in Jan Golinski, "The Chemical Revolution and the Politics of Language," *The Eighteenth Century: Theory and interpretation* 33, (1992), 238–51.
8. Jean-Pierre Poirier, *Lavoisier: Chemist, Biologist, Economist* (Philadelphia: University of Pennsylvania Press, 1996), 188.
9. Cf. François Dagognet, *Tableaux et langages de la chimie: essai sur la representation* (Seyssel: Champ Vallon, 2002).
10. Antoine Lavoisier, Louis-Bernard Guyton-Morveau, Claude Bertholet and Antoine François Fourcroy, *Method of Chymical Nomenclature*, transl. by James St. John (London: G. Kearsley, 1788); Id., *Método de la nueva nomenclatura*

química, transl. by Pedro Gutiérrez Bueno (Madrid: Don Antonio de Sancha, 1788); Id., *Metodo di nomenclatura chimica*, transl. by Pietro Calloud (Venezia: Lorenzo Baseggio, 1790); Id., *Methode der chemischen Nomenclatur für das antiphlogistische System*, transl. by Karl von Meidinger (Vienna: Christian Wappler, 1793).
11. Denis I. Duveen and Herbert S. Klickstein, "The Introduction of Lavoisier's Chemical Nomenclature into America", *Isis* 45, no. 3 (1954), 278–92, here 281–82.
12. Antoine Lavoisier, *Elements of Chemistry. In a New Systematic Order, Containing all the Modern Discoveries*, transl. by Robert Kerr (Edinburgh: Creech, 1790). The word "systematic" was added by the English translator.
13. See Bernadette Bensaude-Vincent and Ferdinando Abbri, eds., *Lavoisier in European Context: Negotiating a New Language for Chemistry* (Nantucket: Science History Publications 1995).
14. See Trevor H. Levere, "Lavoisier: Language, Instruments, and the Chemical Revolution", in *Nature, Experiment, and the Sciences*, eds. Trevor H. Levere and William R. Shea (Dordrecht: Kluwer, 1990), 207–23.
15. Jean-Baptiste le Rond d'Alembert, "Éloge de M. du Marsais," in *Encyclopédie, ou Dictionnaire raisonné des sciences, des arts et des métiers* (1757) quoted after Zofia Florczak, *Europejskie źródła teorii językowych w Polsce na przełomie XVII i XIX wieku: studia z dziejów teorii języka i gramatyki* (Wrocław: Zakład Narodowy im. Ossolińskich, 1978), 107.
16. See the chapter of Miles MacLeod in this volume.
17. Cf. Etienne Bonnot de Condillac, *Œuvres complètes de Condillac.* Vol. 1. *Essai sur l'origine des connaissances humaines* (Paris: Lecointe et Durey, 1822 [1746]), 325–6.
18. M. l'abbé de Condillac, *La logique, ou Les premiers développements de l'art de penser: ouvrage élémentaire* (Paris: L'Esprit et de Bure l'aîné, 1780), 133–4.
19. My translation: "Si les langues sont de véritables instruments que les hommes se sont formés pour faciliter les opérations de leur esprit, il est important que ces instruments soient les meilleurs qu'il est possible". Antoine Lavoisier, "Mémoire sur na nécessite de réformer & de perfectionner la nomenclature de la chimie, lu à l'Assemblée publique de l'Académie Royales des Sciences de 18 Avril 1787", in: idem, Guyton-Morveau, Bertholet and Fourcroy, *Méthode*, 5–25, here 8.
20. Poirier, *Lavoisier*, 189.
21. Lavoisier, *Elements of Chemistry*, 52.
22. *Ibid.*
23. Forming *alkali* (= here *ammoniak*)—see Justis von Liebig et al., *Handwörterbuch der reinen und angewandten Chemie*, Vol. 1 (Braunschweig: Vieweg und Sohn, 1842), 193–4.
24. Particularly Priestley, who defended his phlogiston approach, rejected the nomenclature as it was bound to a theoretical system he did not share. Cf. Jan Golinski, *Making Natural Knowledge. Constructivism and the History of Science* (Chicago and London: Chicago University Press, 2005), 118–9.
25. José Ramón Bertomeu Sánchez, Rosa Muñoz Bello, "Azoote y sulfureto. Debates y propuestas en torno a la terminología química durante la primera mitad del Siglo XIX," *Revista de Investigación Lingüística,* no. 13 (2010): 279–306, esp. 254–62.
26. *Bibliothek der neuesten physisch-chemischen, metallurgischen, technologischen und pharmaceutischen Literatur* 1, (1788): 309–26. The article was in fact a translation of a French review from 1787 of Jean Claude Delamétherie, published in *Journal de physique* 31, (1787): 210–8.

27. Anton Lorenz Lavoisier, *System der antiphlogistischen Chemie*, 2nd ed., transl. by Sigismund Friedrich (Hermbstädt, Berlin and Stettin: Friedrich Nicolai, 1803), 113.
28. Chaptal, Jean-Antoine-Claude, *Anfangsgründe der Chemie*, Vol. 4, transl. by Friedrich Wolff (Königsberg: Nicolovius, 1805), 6.
29. Bedřich Wšemír von Berchtold and Jan Swatopluk Presl, *O přirozenosti rostlin, aneb, Rostlinář, obsahugjcj gednánj o žiwobytj rostlinném pro sebe a z ohledu giných žiwotů podlé stawu nyněgssjho znánj: k rozssjženj přirodnictwj: w potaženj na užitečnost w rolnictwj, hospodářstwj, řemeslech, uměnj i obchodu a w wztahowánj obzwlásstnjm na lekařstwj* (Prague: K dostánj u Karla Wiljma Endersa, 1820). The terms in text are used according to contemporary Czech orthography.
30. Jan Svatopluk Presl, Vojtěch Sedláček, *Lučba, čili, Chemie zkusná* (Prague: Jozefa Fetterlowá z Wildenbrunnu, 1828).
31. Josef Jungmann, *Slownjk česko-německý* (Prague: Knjžecí arcibiskupská knihtiskárna, Josefa wdowa Fetterlowá, 1835–1839).
32. Filip Stanislav Kodym, *Zdravověda čili nejlepší způsob, aby člověk svého života ve zdraví a vesele užiti a k tomu dlouhého věku dosáhnouti mohl* (Prague: Jar. Pospíšil, 1854), 25–6. In some other publications, Kodym called nitrogen *dužík* ("he who gives pulp/flesh", regarding its nutritional value for plants): Idem. *Hospodářský klíč čtení o nejhlavnějších základech umění hospodářského, založeného na vědách přírodních* (Prague: Kat. Jeřábková, 1857), 373.
33. Cf. Jan Janko and Soňa Štrbáňová. "Uplatnění nového českého přírodovědného názvosloví na českých vysokých školách v průběhu 19. století", in: *Místo národnich jazyku ve výuce, vědě a vzdělání v Habsburské monarchii 1867–1918 / Position of National Languages in Education, Educational System and Science of the Habsburg Monarchy 1867–1918*, eds. Harald Binder, Barbora Křivohlavá and Luboš Velek (Prague: Výzkumné centrum pro dějiny vědy, 2003), 297–312, here 301.
34. On this approach to language see Alexander Maxwell, *Choosing Slovakia. Slavic Hungary, the Czechoslovak Language, and Accidental Nationalism* (London: I.B. Tauris, 2009).
35. See the almost identical critique of *nitrogen* in Friedrich Wolff and Juan Mieg; Juan Mieg: *Lecciones elementales de química para uso de los principantes* (Madrid: Catalina Piñuela, 1816), 212–3, quoted in Sánchez, Bello, "Azoote y sulfureto," 259.
36. For an early assessment of his language ideas and some notes on its practice, see Marius Kristensen, "H. C. Ørsted og det danske sprog," *Danske Studier* 1, (1904–1905): 49–64.
37. Sven-Göran Malmgren, "The Lexicon of the Nordic Languages in the 19th Century," in *The Nordic Languages. An International Handbook of the History of the North Germanic Languages*, Vol. 2, eds. Oskar Bandle et al. (Berlin: De Gruyter, 2005), 1443–52, here 1444.
38. See, for example Ørsted, "A Speech at the Opening of the Fifth Meeting of the Scandinavian Naturalists" [1845?], in Idem. *The Soul in Nature: with Supplementary Contributions*, transl. from the German by Leonora and Joanna B. Horner (London: Henry G. Bohn, 1852), 214–23.
39. Oersted, Johannes Christianus, *Tentamen nomenclaturae chemicae omnibus linguis scandinavico- germanicis communis* (Copenhagen: Hafniae, 1814).
40. *H. C. Orsted's Theory of Force: An Unpublished Textbook in Dynamical Chemistry*, eds. and transl. Anja Skaar Jacobsen et al. (Copenhagen: Royal Danish Academy of Sciences and Letters, 2003). For the introduction of new vocabulary: § 146 [Danish: 101–3; English: 253–5] and § 150 [D: 104; E: 256–7].

41. All quotations after H.C. Oersted, "Versuch einer allen skandinavisch-Germanischen Sprachen gemeinschaftlichen chemischen Nomenclatur. Eine Einladungsschreiben zu dem diesjährigen Kopenhager Universitätsfeste zum Andenken der Reformation des Religionswesens und der Universität unter Christian III. lateinisch geschrieben . . . (unter den Augen des Verfassers übersetzt)", *Journal für Chemie und Physik* 12, (1814): 113–54, here 115–9. All translations are mine.
42. *Ibid.*, 119.
43. *Ibid.*, 120.
44. *Ibid.*, 120.
45. *Ibid.*, 121, 122.
46. Italics in original: "Daß die Bennenungen wahrhaft seyn, das heißt, keine falschen Begriffe von Dingen einprägen müssen . . . 2. Der Name muß die Sache bezeichnen, nicht beschreiben, wenn dieß nicht durch die Anzeigung der Elemente derselben geschehen kann. 3. Wo es möglich ist, besonders aber bei der Benennung der Elemente aus denen viele Körper erzeugt werden, muß man sich hüten, daß nicht der Name einer Zusammensetzung als Element in den Namen der Elemente eingehe (wie z.B. bei den Wörtern, *Wasserstoff*, *Vandstof*, Salpeterstoff) . . . 4. . . . die Namen der einfachen Dinge fruchtbar sind, dass heißt so beschaffen, daß aus den einzelnen so viele Benennungen abgeleitet werden können, als aus der bezeichneten Sache andere erzeugt werden. 5. Bei der Bildung der vorzüglichsten Namen muß man darauf sehen, dass die *zusammengesetzten* nicht gar zu lang, oder gar zu schwer auszusprechen werden. 6. Fremde Wörter sind zu verwerfen, *wenn sie sich nicht nach den grammatikalischen Regeln der Muttersprache biegen lassen.*" *Ibid.*, 122–3. Italics in original.
47. *Ibid.*, 126.
48. *Ibid.*, 127.
49. *Ibid.*
50. *Ibid.*
51. I thank Jakob Yngvasson, of the University of Vienna, for informing me about recent Icelandic discussions on the language of chemistry.
52. See Jerzy Biniewicz, *Rozwój polskiej terminologii chemii nieorganicznej* (Opole: Wyższa Szkoła Pedagogiczna, 1992).
53. Śniadecki's book was not the first devoted to chemistry in Polish, as in 1791 Józef Krumlowski translated Jacob Reinhold Spielmann's *Institutiones chemiae praelectionibus, academicis accommodatae* (1766–1767, Polish translation 1791). His nomenclature was, however, regarded as anachronistic and archaic and did not play any role in subsequent developments. Włodzimierz Hubicki, "Studia Chemiczne Jędrzeja Śniadeckiego," in *Rzecz o Jędrzeju Śniadeckim*, ed. Irena Stasiewicz (Warsaw: Wiedza Powszechna, 1970), 29–51, here 48.
54. Jędrzej Śniadecki, *Początki chemii stosownie do teraźniejszego tey umiejętności stanu, dla pożytku uczniów i słuchaczów ułożone y za wzór lekcyi akademickich służyć maiące*, Vol. 1–2 (Wilno: Drukarnia Akademicka, 1800).
55. Alexander Chodkiewicz, "O Słownictstwie Chemiczném," in Id., *Chemiia*, Vol. 3. (Warsaw: Zawadzki & Węcki, 1816), I–XIV.
56. Jędrzej Śniadecki, [Letter to the editor], *Pamiętnik Warszawski* 8 (1817): 384–401; Alexander Chodkiewicz, "Odpowiedź na niektóre uwagi Jędrzeja Śniadeckiego, umieszczone w Pamiętniku Warszawskim na miesiąc sierpień roku bieżącego," *Pamiętnik Warszawski* 9 (1817): 59–77.
57. Jan Oczapowski, "Uwagi o tlenie (oxygenium)," in *Projekt do słownictwa chemicznego* (Warsaw: J. Jaworski, 1853), 61–71.
58. "Aby więc jaknajprędzéj uwolnić się z labiryntu w którym dotąd przewodniczył kwasoród, aby w świątyni tak znakomitej nauki rozjaśnić i utrzymać światło

prawdy, koniecznie wypada przystąpić do wyrugowania z niej nie tylko kwasorodu ale i wszystkich innych wyrazów na *ród* zakończonych". *Ibid.*, 64.

59. "Za pomocą tlenu wszystkie ciała tleć, palić się i w proch obracać się są zdolne, bo wszystko co jest palne łączy się z tlenem . . . tlenie i ukwaszenie ciał są dwa wyrazy oznaczające własność jednego ciała oxygenium". *Ibid.*, 65; Italics and Latin in original.
60. *Ibid.*, 68, fn. 1.
61. *Ibid.*, 69, fn. 3.
62. Teodor Teofil Matecki, *Słownictwo chemiczne polskie* (Poznań: Żupański, 1855), here 10.
63. Karol Lilpop, "Uwagi tyczące się języka chemicznego polskiego," *Biblioteka Warszawska* 3, (1859): 421–36, here 422–3.
64. Emilian Czyrniański, "Niektóre uwagi nad słownictwem chemicznem polskim," *Przyroda i Przemysł* 24, (1857): 190–2 (on the use of *kwasoród* in his handbook of anorganic chemistry from 1857); change to *tlen* occurs in Idem. *Chemia nieorganiczna mniejsza ułożona dla szkół realnych* (Krakow: Drukarnia Uniwersytetu Jagiellońskiego, 1874).
65. See e.g. Johannes Andreas Buchner, "Bemerkungen über System und Kunstsprache der Chemie", *Journal für Chemie und Physik* 8, (1815): 195–224; Johann Salomo Christoph Schweigger, "Anhang von Herausgeber über die chemische Kunstssprache", *Ibid.*, 224–39.
66. Anthony J. La Vopa, "Herder's Publikum: Language, Print, and Sociability in Eighteenth-Century Germany." *Eighteenth-Century Studies* 29, no. 1 (1995): 5–24.

10 Mathematical Machines

Automating Thinking?

Helena Durnová

1. INTRODUCTION

One of the key concepts in computing nowadays is that of a programming language, a name for a tool that facilitates the communication of a task at hand to the computing machine. Meeting initially with some resistance, the use of the term *language* in connection with programming was later scarcely disputed, and indeed quite to the contrary, the metaphor of language invited the use of further metaphors and parallels. Thinking of the induced metaphors, including dialects, syntax, semantics, and pragmatics of programming languages, and comparing these uses with the meaning of those terms in the context of natural languages, raises the question of what natural and programming languages have in common, and even though a modern computer scientist cannot imagine thinking of programming without the notion of a "programming language", there was historically another option: namely, to stick to talking about programming in more technical terms. Studying the choice of the word *language* in the context of computer programming, a rather technical thing, is intrinsically difficult as the word is so widely used that it hardly allows for an explanation for the choice when that choice is not explicitly articulated as a choice by the actors involved. Nofre, Priestley, and Alberts in their recent paper[1] describe how some researchers in the 1950s chose to use the word *language* and others opposed it, but the reasons of either group to choose or not choose the word are not clear. In the following, I attempt to show how the use of this metaphor can be understood in a wider context.

This chapter has four sections, of which the first two serve to set the groundwork for the historical story presented in the last two. In order to allow readers without a background in the history of computing to follow the argument, I have attempted to explain some of the important computing notions in an accessible and brief way, relying on the readers interested in the technical details to be willing to find them in the publications I refer to.

Section 1 recounts some of the linguistically sounding metaphors used in connection with computing in the late 1940s and 1950s in order to give the reader some idea of the vocabulary preceding and surrounding the choice of the metaphor of a "programming language". It points out that *language* was not the only commonplace word used in connection with computers and not even the only one that could seem to be alluding to language: other metaphors with linguistic connotations were (and are) used by programmers, such as *translating* and *interpreting*.

Section 2 deals with the common metaphor of treating mathematics as a language. Since computers were often called mathematical machines, the fact that mathematics was commonly referred to as language may be seen as one possible source of the easy adoption of the word *language* for programming, given programming itself was (and is today) seen as a more technical part of mathematics.

Section 3 looks at the development of the automation of coding in the 1950s by describing the processes of naming that computer scientists employed, including the apparently silent but important choice of the majority of the computing community to suppress the difference between notation and language. The Swiss mathematician Heinz Rutishauser presents an exception to this rule, sticking to the word *notation* well into the 1960s. Although the historical narrative of this chapter revolves around the notion of programming language, the verb *to facilitate* will in fact guide readers through this section. As we will see, the use of terms like *interpreting*, *translating*, and *compiling* in computing preceded that of a programming language.

Section 4 shows how the language metaphor facilitated the use of further linguistic concepts, ultimately leading to the notion of a "programmers' Esperanto". The striving towards a universal programming language in the late 1950s appears a somewhat idealistic activity of the computing people involved in the initiative. However, it is also intriguing that the effort to facilitate communication with the computer by introducing the notion of a programming language resulted in creating another intellectual challenge, namely that of compiler construction.

2. ENTANGLING OF PROGRAMMING AND NATURAL LANGUAGES

> As to the problem of mechanical translation, I frankly am afraid the boundaries of words in different languages are too vague and the emotional and international connotations are too extensive to make any quasimechanical translation scheme helpful.

These are the words Norbert Wiener used in his response to Warren Weaver on April 30, 1947.[2] Wiener's father, Leo Wiener, was in fact a professor of Slavonic languages, which is a connection not at all unusual in the history of computing and abundant among those who were interested in programming languages. The wife of Nikolaus Joachim Lehmann, the East-German specialist on programming languages, was also a linguist.[3] The author of the programming language ALGOL 68, Adriaan van Wijngaarden, developed a special interest in languages in general.[4] When talking about programming languages and the design of compilers, the Austrian computer pioneer and the fourth president of the International Federation for Information Processing (IFIP), Heinz Zemanek,[5] chairman of IFIP Technical Committee 2, Programming Languages, connected those issues to issues of machine translation and brought up the example of Yehoshua Bar-Hillel's sentence "the box is in the pen", thereby illustrating the importance of context in machine translation and his concern with the concept of a language in general. In return, the emergence of programming languages required the development of a distinction between the *natural languages* for languages used for communication between humans and *programming languages*. In his contribution to *Experimenting in Tongues*,[6] Jörg Pflüger extends the parallels between the various formal languages used in computing and natural languages beyond a loose metaphorical connection, almost expecting the theory developed for programming languages to take the same path of development as linguistics. He considers for instance the possibility of applying Saussure's natural language distinction between *langue* and *parole* to programming languages in the future[7] and treating the invention of programming languages as a *linguistic turn*[8] in the developments in computer science.[9]

Still, neither the loose connections with linguists, nor the research on machine translation, can fully justify calling a language something that was before called a system. On the other hand, accepting notions coming from plain language was a common practice among people working on computing with the aid of electronic computers, a new field of activity in the late 1940s and early 1950s. Importing terms from other branches of human activity was one way of naming the new concepts. Approaching the issue of language and computers from that angle, Richard K. Lynch explored the ways in which the words used in computing influenced the way people thought about computers, suggesting that the metaphors used around computers, having originated in fields as diverse as biology and civil engineering, shaped the path for the development of computers.[10] Naming new things in computing, as in other sciences, involved likening new notions to already known ones and communicating new knowledge by making these new concepts accessible through already known notions. This second issue of communicating the advantages of the new thing—the computer—to researchers in other fields is important because computers were seen as a means, not an end. As the Czechoslovak computer pioneer Antonín Svoboda put it in 1952:

> We can say that the production of the machine has by now been finished on the research level to such an extent that we can clearly formulate the code, and the time is ripe for us to make those who are interested acquainted with that code, so that they can anticipate which of their tasks are suitable for a computer and which are not.[11]

Svoboda's words served to warn researchers that the computer itself was not to be seen as panacea: that in order make its use beneficial in research and understand to what kinds of tasks it could be of help, people needed to understand how the computer worked (get acquainted with the code of the computer). The actual use of a computer in research inherently involved the formulation of the research task in a new way, in a way the computer understood, which possibly required even thinking about the research task differently.

As well as exemplifying an effort to mechanize such re-formulations, *mechanical translation* represents a common interest connecting the research of linguists and computer scientists.[12] It is also another instance of the use of computers for a task that is both intellectually demanding and at times boring, and thus translation can be seen, after calculation, as another intellectual activity to be mechanized. Juxtaposing translations between two natural languages and translations from a natural language into some kind of a code or language understandable to the computer reveals the power of the computer as well as its limits. These limits are closely connected to the possibility of reducing human thinking to that which can be communicated through the *language of mathematics*.[13]

3. LINGUA MATHEMATICA

The language of mathematics as a concept is much older than electronic computers. Already Galileo emphasized that God expressed himself in *lingua mathematica*[14] perhaps because many of the people working on programming computers (although they had often originally been trained as physicists, astronomers, engineers, and only sometimes mathematicians) had some formal training in mathematics, which is reflected in their thinking about computing and in anticipating how to assign or "communicate" tasks to computers. Their way of communicating to computers was closely connected with mathematics and with attempts to transcribe the problems in various other branches of research in this "language": that is, by mathematizing them. Before letting the story of communication with computers through programs unfold chronologically, I will focus on the desire expressed by many of those involved in scientific calculations in the mid-1950s that programming languages should be "as close as possible to standard mathematical notation".[15] By focusing on programming as a mathematical activity,[16] I wish to emphasize the fully conscious connection made between mathematics seen as a language and the formulation of means of communication a computer would "understand".

Before applying mathematics, however, one needs to look at the field with different eyes. Mathematization is a necessary condition for applying mathematics to any field.[17] It involves looking at the field with a mathematical mind and isolating those aspects to which mathematical methods can be applied. Physicists have worked on the mathematization of their science for 200 years, but in the immediate postwar years mathematization became widely popular, and researchers in other fields (particularly economics, but also biology and psychology) tried to make their domains calculable.[18] Calculations in themselves, even though intellectually rather demanding, present an auxiliary and rather boring job in astronomy and other branches of science.[19] However, simply explaining the arrival of stored-program computers as due to the desire of delegating the (boring and mistake-prone) task of calculation to a machine misses part of the story. Another part of the story concerns connections between languages, reasoning, and computers. Since natural language is usually vague, one way to enable good reasoning lies in restricting the language we use. Calculations, on the other hand, can be defined as deriving output from input according to certain given rules.[20] Computers function as a tool to use these rules and derive proper conclusions. The connections between the language we speak, good reasoning, and calculations can be traced back to Leibniz, who, besides his proposal of *characteristica universalis*, envisaged how beautiful it would be if all disputes could be resolved by calculations:

> I therefore discovered that there are certain primitive Terms which can be posited if not absolutely, at least relatively to us, and then all the results of reasoning can be determined in numerical fashion . . .
>
> The only way to rectify our reasonings is to make them as tangible as those of the mathematicians, so that we can find our error at a glance, and when there are disputes among persons, we can simply say: Let us calculate, without further ado, to see who is right.[21]

Invention of the computer and computer programming can be regarded as mechanization of thinking per se. In order to understand these developments, one may want to trace the roots of the invention of computers since the beginnings of the twentieth century and the constructivism in mathematics stemming from the late nineteenth-century foundational crisis of mathematics, as William Aspray does.[22] He specifically notes the path of formalism chosen by David Hilbert.

Whether it was the belief that the universe was written in the language of mathematics, or whether the mathematical notation was simply found to be a useful tool in communication with computers in the 1950s, mathematical machines and programming were connected to the language of mathematics. As such the language of mathematics was a major means of "communication" with computers, and several different levels of translation can be discerned therein: namely, mathematization of the problem (which is a translation of a sort), translation of the mathematical formulation into a programming language, and translation (usually denoted by the terms *interpretation* and *compilation*) of a program written in a high-level programming language into the code "understandable" by the computer.

From the very beginning, "communication" with computers was on the agenda of computer pioneers. For example, communicating with ENIAC[23] required physical contact,[24] namely rewiring. Even later, the "message" to the computer had to be coded in a certain way to transform numbers into electrical signals. Coding the information in such a way that it is intelligible by the computer is crucial even nowadays. Such information processing was envisaged by the Czechoslovak computer pioneer Antonín Svoboda, who already in 1950 preferred the term *information processing machines* to the shorter *mathematical machines*, believing that the envisaged machine would perform more than just calculations with numbers.[25] In Czechoslovakia, the two terms, *mathematical machines* and *information processing machines*, were for most purposes used interchangeably.[26]

One of the motivations for making early calculators was the elimination of human error. This held also for the seventeenth-century inventions of Blaise Pascal and Wilhelm Schickard. In the late 1940s it was still thought that the new computing machines would be of help to human computers (i.e., men or women performing the calculations)[27] in the corrections and calculations of mathematical tables.[28] However, as von Neumann asked,[29] how can one expect a machine built from unreliable elements to be reliable? Just like a human being, an unreliable technological element (wire, vacuum tube, etc.) can cause a mistake. This led to replication of the measures employed by human computers, such as repetitions of the same calculations two or three times, in constructing fault-tolerance machines. SAPO (SAmočinný POčítač, i.e., automatic computer) and EPOS (Elektronický počítač střední, i.e., middle-sized electronic computer), the first Czechoslovak computers, are examples of such fault-tolerant machines: the machine itself was to take care of its own failures. For that reason, it had a triple arithmetic unit and a control unit. The triple arithmetic unit meant that all calculations were performed in parallel by three separate units. The results were then compared by the control unit, which took over the decision-making process. If at least two of the three results were the same, the calculation continued; if not, the calculation was repeated.

4. OVERCOMING THE DIFFICULTY OF CODING: AUTOMATING CODING

As has been shown in the previous section, the relatively simple mental tasks of calculating and comparing were thus mechanized in the complex structure of the computer. However, there was still a rather boring mental task to be performed: coding the calculation in such a way that the computer would understand it. Coding has a connotation of inferiority to it. While inventing or discovering mathematical methods for solving a problem was considered smart, its implementation on a computer was considered less intellectually demanding in the same way that computation was considered less skilled work than mathematics. Thus, the best human computers were hired in June 1945 as coders for ENIAC, and since human computers at that time were often women, these coders were then known as "ENIAC girls".[30] Coding, alas, was also soon discovered to be prone to mistakes.

With regard to the solution of tasks with the aid of a computer program, mathematics, mathematical formulation, mathematical notation, and, last but not least, the mathematical training of the programmer were all thought to play a role. For example, in the late 1940s, such ideas were developed simultaneously and most probably independently in the US and in (West) Germany.[31] In 1947, John von Neumann and Herman H. Goldstine claimed that after the problem has been understood, the rest of the preparation of the program could be performed in a "routine manner" by a mathematician or an "even moderately mathematically trained person".[32] Still, in comparing a computer with a brain, John von Neumann made a clear distinction between the language of the brain and the language of mathematics.[33]

The connection between mathematics and computers was firmly established by 1950, but this was not the case with language and computers. To choose a well-known example, in 1948, Konrad Zuse presented his Plankalkül (which is now considered by some to be one of the first instances of something we now refer to as a programming language) as a means for *formulating* tasks.[34] Zuse himself, however, did not refer to language, but a plan of calculations (Plankalkül). Similarly, in 1954 FORTRAN was developed as a FORmula TRANslator primarily as a programming system.[35]

In 1955, a world-wide conference on the development of computers was held in Darmstadt. This meeting was of key importance in the history of computing because it provided the possibility for the scientists from the US, Western Europe, and the Soviet bloc to meet and discuss the details of computer construction and computer programming.[36] At this conference, the Swiss computing specialist Heinz Rutishauser declared the mathematical formulation to be the first step in the description of a method for the simplification of programming for the Z4 computer designed by Zuse.[37] Rutishauser went even further in a discussion to another contribution at the same conference and advocated the rise of automatic programming as a specific mathematical tool: "Automatic programming should become a mathematical tool, just like matrix algebra and integration".[38] Rutishauser called for *algorithmische Schreibweise*, that is, *algorithmic notation*, which soon inspired Hermann Bottenbruch to develop the concept of *algorithmic language*.[39]

There is, however, an instance of a rather early use of the word *language* in connection with computing machine, and that is *machine language* (not a programming language). The use of the term *machine language* seems to have been stimulated by the use of the term *language* in symbolic logic. Similarly, the word *coding* had already been used for the process of sending a message: the original message would be encoded (e.g., a number would stand for each word in the message), the result would then be transmitted, and the receiver would obtain the original message by the inverse process, "decoding". This process is different from encrypting and deciphering, as the concern with coding is not secrecy, but the possibility of transmission: for example, transmitting numbers may be easier than transmitting words. For scientific calculations performed on a computer, the existing mathematical method must be broken down in minute detail, and only in such a form can the detailed instructions be given to the computer. Also, computers need to receive instructions in a form they understand. In the case of electronic computers this would be done by sending impulses through wires, sometimes referred to as "zeroes and ones". Even without knowing anything about programming computers and mathematical methods, it is clear that translating directly into "zeroes and ones" would be a tedious business. To make the task more manageable, computing people developed various methods, ranging from designing instruction sequences and instruction nets by using assembly languages, but also translating, interpreting, and programming languages. While some of those terms seem to refer to technological practice, others seem to make a more direct connection to the art of translation.

The term *instruction sequence* belongs to the technological denotations. It was used in Czechoslovakia in the early 1950s to describe the communication of a problem to the computer in the form of linearly ordered commands.[40] Typing such a straightforward sequence might be clumsy when parts of the calculation are to be repeated, or if the next instruction depends on the result of the previous operation. As such, the concept of "instruction nets" were developed:[41] this new concept played with the idea of branching and loops: rather than proceeding in linear fashion, the computation scheme would return to perform technically the same operations with different numbers (loops) or would allow for decisions between two different operations (branching). All this was based on the results of the previous calculations. Unlike the notion of sequence, the notion of net allowed for the branching and loops, and thus opting for the new term *instruction net* reflected the need to describe a new concept here. On the other hand, this new term was confusing as well: the report from 1952 explicitly mentions EDSAC (Electronic Delay Storage Automatic Calculator), a computer whose designers used the word *program* with a meaning equivalent to "instruction net".[42] Indeed, although the word *program* was also used elsewhere to denote the same concept,[43] the Czechoslovak computer pioneer Antonín Svoboda opted for the more descriptive *instruction net*, thus obscuring the connection between the two terms. This even led to bizarre formulations like the following one: "People who design instruction nets are called programmers".[44]

Writing down the instruction sequence for the computer also involved references to the specific places in the memory of the computer. These places are called *addresses*. The computer would *write* on these spaces and *read* from them. From the point of view of instruction nets, the number of addresses of the specific computer was significant, since it can in general be claimed that the more addresses the computer had, the shorter the instruction sequence and the simpler the instruction net. For this reason, SAPO, the first Czechoslovak computer, had five addresses. The machine was complex, but writing the instruction nets was so easy that no further automation of the process was needed.[45] One aid for writing programs was frequently used: namely, putting the program together from blocks, called subroutines (in Czech *instrukční podsítě* / *instruction subnets*), that denoted descriptions of those tasks that are performed very often. These routines were stored in a library, and the program would *assemble* them. However, one-address computers soon prevailed, as their construction allowed for a much faster performance. As a consequence, the method of programming was not a comfortable one because programs for one-address computers needed to be closely tied to the process of moving the numbers around the computer so the desired operation could be performed. These programs are not as close to the ordinary mathematical language description as programs for five-address computers. Keeping track of where the program should write and read the correct values—keeping track of the right addresses—was a demanding, but not an intellectually rewarding, job. This rather dull part had to be assigned to the computer. These intermediate methods were called *compiling* and *interpreting* and can be regarded as instances of translation.

Prior to electronic digital computers, calculations were performed by humans, and this involved a lot of tacit knowledge.[46] Approximation could even influence the choice of the next step, for example, whether the calculation should be stopped when the result is zero. The computers needed to be told all this and needed to receive this information in a different format from that of a physicist, astronomer, or any other researcher who needs to use the particular scientific calculation. With hindsight, it might be said that the difficulty of translating methods used by humans into methods used by computers was underestimated.[47] In the late 1950s, the distinction between the intellectual and routine part of the programming process became apparent, and just as there were coders for ENIAC, the split between the intellectual and the routine part of the job among humans was envisaged: the intellectual part would be assigned to *programmers*, the routine to (human) *compilers*.[48] Programming was thus reserved for humans, while the compilation was soon to be automated.

Mechanizing the routine part of programming initially adopted different names in different countries and communities. While the Russians spoke of a programming program,[49] the British spoke of autocoding,[50] and the Americans of automatic programming.[51] Connected to translation (probably in analogy to translating between natural languages), the terms *interpreting* and *translating* were used for the systems performing mechanized translation of the program simultaneously and consecutively, respectively. In this sense, simultaneous translation, or interpretation, would mean translating one instruction and immediately performing it, while consecutive translation, or translation, would mean translating the program as a whole and then performing it. It is worth noticing that translators and interpreters do the boring task, just as mechanical translations between natural languages were designed not for translating poetry, but technical and scientific texts. By contrast, the tool for the (human, highly intelligent) programmers was eventually called *language*, or more specifically, *programming language*.[52]

5. ALGOL AS PROGRAMMERS' ESPERANTO

Programming languages have by now become a commonplace term for describing a tool for communicating a task to a computer, although the idea of calling this tool a language was not embraced immediately.[53] The efforts to define a universal programming method were motivated by enabling communication of programs between humans as well as transferring the program from one computer to another. I will show this using the example of the programming language ALGOL 60, a language that "changed the face of current programming"[54] and whose intellectual content motivated people to call it "a programmers' Esperanto",[55] or, to coin a phrase, "ALGOL-like languages".[56]

ALGOL 60 stands out in terms of its creation. In the late 1950s, small groups of scientists in two societies, the Association for Computing Machinery and *Gesellschaft für andgewandte Mathematik und Mechanik*, decided to form a joint committee that was to design a universal programming method.[57] This committee decided to call the first draft of the language IAL, for International Algebraic Language, but this name was soon changed to ALGOL, for ALGOrithmic Language. The abbreviation ALGOL was probably chosen because a few of the people designing the language were originally astronomers and Algol is a star in the constellation of Perseus,[58] but for some it also reflected a change of focus from algebra to algorithms.

Beginning in the late 1950s as a German-US centred initiative, the effort to design a programming language spread to the Netherlands, England, France, Denmark, and even to Soviet Union and the Eastern bloc, through personal invitations,[59] the publication of the preliminary (1959) report[60] and later (1960) the "Report on the Algorithmic Language"[61] in scientific journals, and through the *ALGOL Bulletin*.[62] Despite the initial enthusiasm, ALGOL 60's aspiration to become a universal language vanished within a decade after the publication of the report in 1960. Unlike an earlier programming method, FORTRAN (FORmula TRANslator), whose new and improved versions are still being issued, ALGOL 60 exists rather in the form of an elaborate model (of a programming language). Moreover, contrary to the effort to unite programming languages, it is nowadays often thought more productive to design a new programming language, tailored for the particular task at hand, rather than a universal programming language.

The parallel to artificial language raises an interesting question: what role were the earlier conceptions of "artificial language" playing in the attempts to institute universal programming languages? Were programming languages bound to embark upon the same pattern as artificial languages for communication among humans from the start? In the context of natural language, thoughts on universal language and attempts to restore the condition prior to the Tower of Babel have recurred throughout history. Apart from the efforts of Wilkins, Dalgarno, and Comenius, we can find one arguably successful candidate, Zamenhof's Esperanto, which was designed in the late nineteenth century. Just how successful this language became is reflected in the widespread metaphorical use (among programmers) of its name, Esperanto, in both the positive and negative sense.[63] Becoming a programmers' Esperanto in the positive sense,[64] ALGOL 60 for some time indeed fulfilled the task of communicating algorithms between programmers. This was a result of the decision of the Association for Computing Machinery to publish algorithms in ALGOL 60. The same decision was also accepted by the Czechoslovak journal *Applications of Mathematics* in 1965. ALGOL 60 was also one of the first languages taught to programmers, and even though some might have later had doubts about ALGOL 60's usefulness, almost every programmer knew the language. ALGOL 60 also encouraged research on the properties of formal languages, mostly drawing on the types of grammars defined by Noam Chomsky.[65]

ALGOL 60 failed on the level of becoming an easy tool for transferring a program from one computer to another.[66] To enable the use of ALGOL 60, each computer had to have a compiler to be able to execute programs written in ALGOL 60. Thus, the intellectual endeavour was moved another step up. It was then relatively easy to write programs in ALGOL 60, but designing compilers was considered an intellectual challenge.[67] On this level different groups chose different subsets of ALGOL 60 and different aspects of the language to be included in their particular version.[68] These restrictions made it impossible to directly transfer the program from one computer to any other, which had been one of the major goals of the project when it started.

Compilers as such present an interesting point in the automation of programming. A compiler may check the syntax of the program written by the programmer, but it cannot always ensure the program will compute what it was designed to compute. They presented another step in the automation of programming, bringing a new level of complexity to relationship between program and programmer.

6. CONCLUSION

Language is a commonplace word whose metaphorical usage in programming automatic computers stimulated thinking about computers as entities to which a task can be communicated. It was not the only linguistic metaphor used in connection with computers, the other metaphors being translating, interpreting, and compiling. The process of translation of the task into machine language by a human was considered difficult, and methods were sought to make it easier for human computers. The concept of a programming language, being close to mathematics as most computer programmers understood, was meant to make things easier. It would be even easier to have only one programming language, common to all programmers. Interestingly, the practical demand for a unified notation was easily moulded into a dream of a universal language. However, difficulties emerged on the level of compiler constructions, reminiscent of machine translation. After automating and simplifying some aspects of programming computers, there still remained something for humans to solve. Making use of the (now omnipresent) language metaphor in programming, the parallel with machine translation allows for a metaphorical explanation of what cannot be automated. Perhaps the thinking that cannot be described mathematically, and hence cannot be processed by the computer, is of the same kind as the aspects of language that defy mechanical translation.

NOTES

1. David Nofre, Mark Priestley, and Gerard Alberts, "When Technology Became Language: The Origins of the Linguistic Conception of Computer Programming, 1950–1960," *Technology and Culture* 55, no. 1 (2014): 40–75.
2. Warren Weaver, "Translation," in *Machine Translation of Languages*, eds. William N. Locke and A. Donald Booth (New York: The Technology Press of the MIT and John Wiley and Sons; London: Chapman and Hall, 1955), 17. This is a reprint of Warren Weaver's talk of July 15, 1949.
3. Hartmut Petzold, "Nikolaus Joachim Lehmann from Dresden and the international ALGOL project," talk presented at the conference "Pioneering Software in the 1960s in Germany, The Netherlands, and Belgium", held in Amsterdam, November 2, 2006.
4. Gerard Alberts, "Van Wijngaarden and ALGOL 68," talk presented at the conference "Pioneering Software in the 1960s in Germany, The Netherlands, and Belgium", held in Amsterdam, November 2, 2006.
5. Private communication with Heinz Zemanek, June 13, 2008, Vienna.
6. Jörg Pflüger, "Language in Computing," in *Experimenting in Tongues*, ed. Matthias Dörries (Stanford: Stanford University Press, 2002), 125–162.
7. Pflüger, 145.
8. For the use of the notion of linguistic turn in philosophy, cf. Richard Rorty, "Metaphysical Difficulties of Linguistic Philosophy," in *The Linguistic Turn*, ed. Richard Rorty (Chicago, London: The University of Chicago Press, 1967), 8.
9. Pflüger, 156.
10. Richard K. Lynch, *From Analytical "Engines" to Data "Architectures" & Software "Engineers": How the Words we Used Influenced our Thinking about Computing*. Manuscript, dated January 1995, CBI, collection MSM papers, box 54, folder SWE—varia.
11. Antonín Svoboda, "Z pracovní konference pořádané oddělením strojů na zpracování informací při Matematickém ústavu Československé akademie věd," *Časopis pro pěstování matematiky* 78, no. 1 (1953), 28.
12. Janet Martin-Nielsen, "'It Was All Connected': Computers and Linguistics in Early Cold War America," in *The Construction of Social Science in Cold War*

America, eds. Mark Solovey and Hamilton Cravens (New York, NY: Palgrave, 2012), 53–78.

13. Following the article by A. M. Turing (Alan M. Turing, "Computing Machinery and Intelligence", *Mind* (New Series) 59, no. 236 (1950), 433–460) devoted to the question of whether machines can think, this could be further linked to artificial intelligence, but that would be beyond the scope of this paper.
14. See e.g. Matthias Dörries, "Language as a Tool in the Sciences," in *Experimenting in Tongues*, ed. Matthias Dörries (Stanford: Stanford University Press, 2002), 10.
15. Heinz Rutishauser, *Description of ALGOL 60. Handbook for Automatic Computation, Vol. I, Part A.* (Berlin and New York: Springer Verlag, 1967), 6.
16. Programming conceived of as a mathematical activity is different from programming as tinkering (i.e., making the machine do what one wants to do without knowing precisely what one is doing); cf. e.g. Robert S. Boyer and J. Strother Moore, "Preface," in *The Correctness Problem in Computer Science*, eds. Robert S. Boyer and J. Strother Moore (London: Academic Press, 1981), vii.
17. For a detailed account, see Gerard Alberts, *Jaren van berekening: Toepassingsgerichte initiatieven in de Nederlandse wiskunde-beoefening 1945–1960* (Amsterdam: Amsterdam University Press, 1998).
18. Gerard Alberts, "Mathematization and Mathematical Modelling," invited talk at *Colloquium SVSHM*, Prague, January 23, 2010.
19. For more about human computers, see David Alan Grier, *When Computers Were Human* (Princeton: Princeton University Press, 2005).
20. Cf. e.g. Konrad Zuse, "Über den Allgemeinen Plankalkül als Mittel zur Formulierung schematisch-kombinativer Aufgaben," *Archiv der Mathematik* 1, no. 6 (1948), 441–449.
21. Philip Wiener, *Leibniz: Selections* (New York: Charles Scribner's Sons, 1951), 51.
22. William F. Aspray, *From Mathematical Constructivity to Computer Science.* Unpublished PhD dissertation, University of Wisconsin, Madison, 1980.
23. ENIAC stands for Electronic Numerical Integrator and Computer. It was not a stored-program computer, but its construction is considered an important landmark in the history of computing.
24. Liesbeth de Mol, "How to Talk with a Computer. An Essay on Computability and Man-Computer Conversations," in *Off Topic: Übersetzen. Zeitsschrift für Medienkunst der KHM*, ed. S. Stallschus, S. Schönberg, C. König (Köln: Verlag der Kunsthochschule für Medien Köln, 2008): 80–89.
25. "Aby nebolela hlava (Against a headache)", script for a TV show broadcasted on March 14, 1957. Masarykův ústav—Archiv Akademie věd ČR (Masaryk Institute and the Archive of the Academy of the Sciences of Czech Republic), minutes of the presidium meeting, June 19, 1957.
26. Jiří Klir, *Matematicke stroje* (Prague: Práce, 1961), 5.
27. Human computers were people whose job was performing long calculations on paper. The results of their work were used by physicists, astronomers, and other scientists and engineers.
28. David Alan Grier, *When Computers Were Human* (Princeton: Princeton University Press, 2005).
29. John von Neuman (1903–1957) is one of the key figures in history of computing. The standard architecture of the computer is called von Neumann architecture.
30. Nathan Ensmenger, *The Computer Boys Take Over: Computers, Programmers, and the Politics of Technical Expertise* (Cambridge, MA: MIT Press, 2010), 35. See also Janet Abbate, *Recoding gender* (Cambridge, MA: MIT

Press, 2012) and Thomas J. Misa, ed. *Gender Codes: Why Women Are Leaving Computing* (IEEE-CS Press/Wiley, 2010).

31. Zuse, 441.
32. John von Neumann and Herman H. Goldstine, *Planning and Coding of Problems for an Electronic Computing Instrument. Report on the Mathematical and Logical Aspects of an Electronic Computing Instrument, Part II, Volume I* (Princeton: Institute for Advanced Study, 1947), 20.
33. John von Neumann, *The Computer and the Brain* (New Haven: Yale University Press, 1958), 80–82.
34. Zuse, 441–442.
35. John Backus, "History of FORTRAN I, II, and III, " in *History of Programming Languages I*, ed. Richard L. Wexelblat (New York, NY: ACM, 1978), 25–74.
36. Hartmut Petzold, "Eine Informatiktagung vor der Gründung der Informatik. Die Darmstädter Konferenz von 1955," in *Form, Zahl, Ordnung. Studien zur Wissenschafts- und Technikgeschichte. Ivo Schneider zum 65. Geburtstag*, ed. Rudolf Seising, Boethius, Band 48 (Stuttgart: Steiner, 2004), 759–782.
37. Heinz Rutishauser, "Massnahmen zur Vereinfachung des Programmierens," in *Elektronische Rechenmaschinen und Informationsverarbeitung*, ed. Johanness Wosnik (Braunschweig: Friedr. Vieweg & Sohn, 1956), 26.
38. Original German formulation: "das automatische Programmieren sollte zu einem mathematischen Werkzeug werden, genau so wie die Matrizenrechnung und die Integration". Heinz Rutishauser, contribution to the discussion after the contribution of N. Joachim Lehmann, "Bemerkungen zur automatisierung der Programmfertigung für Rechenautomaten", in *Elektronische Rechenmaschinen und Informationsverarbeitung, NTF* [Nachrichtentechnische Fachberichte], supplement of NTZ, Vol. 4, ed. Johannes Wosnik (Braunschweig: Friedr. Vieweg & Sohn, 1956), 143.
39. For initial discussion of this development, see Helena Durnová and Gerard Alberts, "Was ALGOL 60 the First Algorithmic Language?," *IEEE Annals of the History of Computing* 36, no. 4 (2014): 104–106.
40. Helena Durnová, "Embracing the Algol Effort in Czechoslovakia," *IEEE Annals of the History of Computing* 36, no. 4 (2014): 26–37.
41. Research in instruction networks (Výzkum instrukčních sítí pro samočinný počítač). Research report of the Central Institute for Mathematics, Department for Information Processing Machines (Závěrečná zpráva Ústředního ústavu matematického, oddělení strojů na zpracování informací) no. 7 (1952), Archive of the National Technical Museum in Prague (Archiv Národního technického muzea v Praze).
42. Maurice V. Wilkes, David J. Wheeler, and Stanley Gill, *The Preparation of Programs for an Electronic Digital Computer. With Special Reference to the EDSAC and the Use of a Library of Subroutines* (Cambridge, MA: Addison-Wesley, 1951).
43. H. Ruthishauser, A. Speiser, and E. Stiefel, "Programmgesteuerte digitale Rechengeräte (elektronische Rechenmaschinen)," in *Mitteilungen aus dem Institut für angewandte Mathematik, no. 2* (Basel: Birkhäuser, 1951).
44. Klír, 31.
45. Jiří Raichl, "The First Year of Operation of the Digital Computer SAPO," *Information Processing Machines* 7 (1959), 143–146.
46. This perspective is due to Michael S. Mahoney.
47. This underestimation is reflected upon e.g. in Liesbeth De Mol, Maarten Bullynck, and Martin Carlé, "Haskell before Haskell: Curry's Contribution to Programming (1946–1950)," in *Proceedings of the Programs, Proofs, Process and 6th International Conference on Computability in Europe* (CiE'10),

eds. Fernando Ferreira, Benedikt Löwe, Elvira Mayordomo, and Luís Mendes Gomes (Berlin, Heidelberg: Springer-Verlag), 108–117.

48. Klír, 27.
49. Andrei Petrovich Ershov, *Programmiruščaja programma dlja BESM* (Moscow: Izdatelstvo Akademii Nauk SSSR, 1958).
50. R. A. Brooker, "The Autocode Programs Developed for the Manchester University Computer," *The Computer Journal* 1, no. 1 (1958): 15–21; R. A. Brooker, "The Programming Strategy Used with the Manchester University Mark I Computer," *Proceedings of the IEE, Part B: Radio and Electrical Engineering* 103, no. 1 (1956), 154.
51. Note e.g. the title of the conference *Symposium on Automatic Programming for Digital Computers*, Sponsored by the Navy Mathematical Computing Advisory Panel at Washington, D.C., May 1954.
52. For a detailed discussion of this development, see Nofre, Priestley, and Alberts, 2014.
53. Gerard Alberts and Edgar G. Daylight, "Universality Versus Locality: The Amsterdam Style of ALGOL Implementation," *IEEE Annals of the History of Computing* 36, no. 4 (2014): 52–63.
54. Robert W. Bemer, "A Politico-Social History of ALGOL," in *Annual Review of Automatic Programming* 5, ed. Richard Goodman (Oxford: Pergamon Press, 1969), 151.
55. In the 1960s, Antonín Svoboda called the programming language ALGOL a "programmers' esperanto" in his regular lecture on mathematical machines for students of numerical mathematics.
56. Karel Čulík, "On Equivalent and Similar Grammars of Algol-like Languages (Preliminary communication)," *Commentationes Mathematicae Universitatis Carolinae* 5, no. 2 (1964): 93–95; John C. Reynolds, *Theories of Programming Languages* (Cambridge: Cambridge University Press, 1998), 415–446.
57. For an overview of the early history, see David Nofre, "Unraveling Algol: US, Europe, and the Creation of a Programming Language," *IEEE Annals of the History of Computing* 32, no. 2 (2010): 58–68.
58. Richard L. Wexelblat, ed. *History of Programming Languages* (New York: Academic Press, 1981), 696.
59. Richard W. Hamming, President of ACM, and Robert Sauer, President of GAMM, letter to A. A. Dorodnitsyn, August 10, 1959. Ershov archive (online at http://ershov.iis.nsk.su/english), folder 135, list 314.
60. Alan Perlis and Klaus Samelson, "Preliminary Report—International Algebraic Language," *Communications of the ACM* 1, no. 12 (1958): 8–22; Alan Perlis and Klaus Samelson, "Report on Algorithmic Language ALGOL by the ACM Committee on Programming Languages and the GAMM Committee on Programming," *Numerische Mathematik* 1 (1959): 41–60. Despite the different titles, the content of the two papers is almost identical.
61. John W. Backus, "Report on the Algorithmic Language ALGOL 60," *Communications of the ACM*, 3, no. 5 (1960): 299–314; John W. Backus, "Report on the Algorithmic Language ALGOL 60," *Numerische Mathematik* 2 (1960): 106–136.
62. *ALGOL Bulletin* was edited by Peter Naur and distributed to computing centres in Europe. It was published since March 1959.
63. David Nofre in personal communication pointed out that the negative use of Esperanto is far more widespread.
64. For the study of common traits of Esperanto and ALGOL 60, see Federico Gobbo and Helena Durnová, "From Universal to Programming Languages," *Informal Proceedings of the Computability in Europe conference*, 2014. (10 pages, published electronically / on a CD.)

65. Karel Čulík, "Semantics and Translation of Grammars and ALGOL-like Languages," *Kybernetika* 1 no. 1 (1965): 47–49; Jozef Gruska, "Structural Unambiguity of ALGOL MOD," *Commentationes Mathematicae Universitatis Carolinae* 6, no. 3 (1965): 281–327; Jiří Kopřiva, "A Note on the Structure of Certain Predicates Concerning the Sublanguages of ALGOL 60," *Kybernetika* 1, no. 2 (1965): 122–126.
66. David Nofre and Gerard Alberts, "Localizing ALGOL: Industrial, Academic, and Military Sites of Early Software Production in the US and Western-Europe (1958–1965)," unpublished manuscript, presented at Software for Europe workshop, Lorentz Center, Leiden, 9–13 September 2010.
67. On compilers as an intellectual challenge see e.g. Brian Randell, "Whetsone Algol Revisited or Confessions of a Compiler Writer," paper presented at the First Compiler Writing Seminar, May 20, 1964, Brighton College of Technology; Brian Randell, "Reminiscences of Whetstone ALGOL," *University of Newcastle upon Tyne, Computing Science, Technical Report Series*, No. CS-TR-1190, p. 3.
68. Alberts and Daylight.

Part 3

Imagining Universal Languages

11 Seventeenth-Century British Projects for a Universal Language and Their Reception in the Augustan Age

The Cases of John Wilkins and Jonathan Swift

Rocío G. Sumillera

1. DEVISING UNIVERSAL OR "REAL" CHARACTERS AND PHILOSOPHICAL LANGUAGES IN THE SEVENTEENTH CENTURY

In November 1662 Charles II chartered the Royal Society of London for the Improvement of Natural Knowledge. It was a time of major political change, with the monarchy itself only just restored, and of scientific change, with a move towards empirical experiments and the direct observation of nature replacing the scholastic tradition and its understanding that scientific truths were based on the opinions of ancient and medieval authorities (chiefly Aristotle) and religious doctrine.[1] This new orientation of the pillars of scientific study, together with new discoveries and inventions, had an impact on all fields of thought and spurred the literary and artistic imagination. Research by members of the Royal Society caught the attention of many seventeenth- and eighteenth-century writers, who either supported such investigations or overtly ridiculed them in their works. Abraham Cowley, author of "A Proposition for the Advancement of the Experimental Philosophy" (1661) and the poem "To the Royal Society", prefixed to Thomas Sprat's *History of the Royal Society of London* (1667), is representative of the supporters, while Jonathan Swift and Henry Fielding were notable detractors.[2] As raw material for their parodies, writers such as Swift and Fielding turned to the *Philosophical Transactions of the Royal Society*, which was the first scientific journal of the English-speaking world, the official testimony of the meetings and experiments of the Society, and a publication "widely read" by scientists and "well-read laymen generally".[3]

The shift in the appreciation of science ran in parallel with projects to devise a universal and philosophical language that would stimulate and facilitate scientific advancement. The seventeenth century witnessed, on the one hand, the loss of the privileged position that for centuries Latin had enjoyed as Europe's international language and, on the other, the craze for developing a universal language, encouraged partly by the exponential expansion of trade and the increase of international dialogue among scholars, scientists, and research institutions such as the Royal Society.[4] The benefits of a universal language appeared evident: by eradicating the problems derived from the multiplicity of tongues, international communication would improve, and even at a domestic scale this would solve the issue of the "'Babellish confusion' of Renaissance English".[5] The quest for a universal language began by seeking a universal writing system: the so-called "universal" or "real character", which would signify things, unlike "vocal character" (i.e., alphabetical writing), which represented spoken sounds. Underlying the notion of the real and universal character "there is the assumption, ultimately deriving from Aristotle that there is a set of concepts which are universally shared by all mankind, and which are identifiable independently of any particular form in which they may be expressed".[6] As the Dutch Jesuit Hermann Hugo argued in *De prima scribendi origine et universa rei literariae antiquitate* (1617), men "receive the same concepts of things through the senses, and so the differences that arise among them must result from the terms that are applied to these concepts rather than from the concepts themselves".[7] Consequently, if language merely encodes these universal concepts (which follow from the unity of human reason), an agreement on how to encrypt them graphically would lead to a universal character readable in all languages.

The internationality of astronomical symbols and the languages of music, chemistry, and mathematics served as inspiration to seventeenth-century scientists with linguistic concerns, many of whom were, not coincidentally, mathematicians (Wilkins, Seth Ward, Newton, Mersenne, Descartes, Leibniz). Numbers became a model for universal language projects for other reasons: even if "there were only ten symbols to be memorised, all concepts of quantity could be expressed by them without any limitation and without any ambiguity, number being an absolute for all speakers"; plus, "the fixed order in which they occurred made their signification extremely easy to recall".[8] Robert Boyle, who would subsequently become sceptical about the universal character project, wrote to Samuel Hartlib in 1647, saying that, just as "arithmetical characters are understood by all the nations of Europe", "I conceive no impossibility that opposes the doing that in words".[9]

In addition, Egyptian hieroglyphs were at this time thought to represent things,[10] and in the late sixteenth century the word spread from returning missionaries and travelers from China that Chinese characters "represented things directly, rather than representing them indirectly through spoken sounds".[11] The first to discuss the notion of universal character in England was Sir Francis Bacon, who, in his *Advancement of Learning* (1605), claimed the following in relation to Chinese characters:

> In China and the provinces of the furthest East there are in use at this day certain *real characters*, not nominal; characters, I mean, which

> represent neither letters nor words, but things and notions; insomuch that a number of nations whose languages are altogether different, but who agree in the use of such characters . . . communicate with each other in writing; to an extent that any book written in characters of this kind can be read off by each nation in their own language.[12]

Bacon regarded as imperfections of language words for things that do not exist, the arbitrary assignment of names to things, ambiguity of meaning, metaphorical language, phraseology, grammatical irregularities and exceptions, synonymy, and the mismatch between orthography and pronunciation.[13]

The fascination with emblems, anagrams, and cryptography, or secret-writing, in sixteenth- and seventeenth-century Europe becomes apparent in works such as Tritheim's *Polygraphiae libri sex* (1518), Baptista Porta's *De furtivis literarum notis* (1563), and Gustav Selenus's *Cryptomenytices et cryptographiae* (1624). Furthermore, "brachigraphy", or shorthand, was a popular British phenomenon, as can be seen in Timothy Bright's *Characterie* (1588), John Willis's *Art of Stenographie* (1602) and Noah Bridges's *Stenography and Cryptography or The Arts of Short and Secret Writing* (1659).[14] Secret communication was also the focus of John Wilkins's *Mercury, or The Secret and Swift Messenger* (1641), which includes a chapter on universal character in which Wilkins wishes there existed "an universal character to express Things and Notions, as might be legible to all People and Countries, so that Men of several Nations might with the same ease both write and read it".[15] Wilkins argued that such a universal character would naturally "conduce to the spreading and promoting of all Arts and Sciences".[16] Another influence on the devising of a universal language came in the form of works dealing with teaching the deaf and dumb to speak, which were mainly concerned with speech training and gesture. Notable publications here include John Bulwer's *Chirologia* (1644), Meric Casaubon's *A Treatise Concerning Enthusiasme* (1656), George Sibscota's *The Deaf and Dumb Man's Discourse* (1670), and Daniel Foote's translation into English of J.C. Amman's *Surdus loquens*, entitled *The Talking Deaf Man* (1693). The Scottish schoolteacher George Dalgarno contributed to this field with his *Didascalocophus or The Deaf and Dumb Mans Tutor, to Which Is Added A Discourse of the Nature and Number of Double Consonants* (1680), a study of phonotactics.[17]

Francis Lodwick, a merchant of Flemish origin based in London, was the author of three of the most remarkable schemes for universal writing of the seventeenth century: *A Common Writing, Whereby Two, although Not Understanding One the Other's Language, yet by the Helpe Thereof, May Communicate Their Minds One to Another* (1647), *The Groundwork, or Foundation Laid (or so Intended) for the Framing of a New Perfect Language: and an Universal* or *Common Writing* (1652), and "An Essay towards a Universal Alphabet and a Universal Primer", published in the *Philosophical Transactions of the Royal Society* in June 1686.[18] Other works on the subject include Cave Beck's *The Universal Character* (1657), devised as a numerical dictionary,[19] and Athanasius Kircher's *Polygraphia nova et universalis* (1663), with lists of numbered words in several languages to facilitate translation. *Technica curiosa* (1664) by Caspar Schott describes the work of Pedro Bermudo, a Spanish Jesuit who wrote *Arithmeticus nomenclator* circa 1653; since his proposal was to replace words by numbers, he identified forty-five classes of the most frequently mentioned things and assigned a number to each.[20] Schott also comments on *Character, pro notitia linguarum universali* (1661), by the German scholar Johann Joachim Becher, who compiled a 10,000-word dictionary in Latin that provided basic characters made up of points and strokes for a list of terms.[21]

Ideally a universal language would be rational and philosophical enough to connect ideas and words in an unequivocal and logical way and would be evident to everyone irrespective their mother tongue. Ultimately it would be a far more accurate vehicle for the representation of knowledge and reality than natural languages.[22] The ideal of the philosophical language was to univocally match words with things in such a way that words would indicate the nature of the things themselves. In a philosophical language the number of names would match the number of objects, and each name would represent the properties of its associated object. Thus, such a language would work as an algebraic formula "in which the orthographic constituents of each word could denote the semantic features of the word", and so, for instance, "the graphemes of the word *tree* would indicate the characteristic qualities of a tree such as *with roots*, *leaf bearing*, and the like".[23] The first proposals for a philosophical language begin in 1629, with Descartes's letter to Marin Mersenne, who discussed the idea in *Harmonie universelle* (1636), where he listed the features that, from his viewpoint, would qualify a language as excellent. Descartes suggested "the creation of a language based upon the ordering of simple ideas", which would nonetheless "presuppose a perfect analysis of knowledge".[24] Through the influence of Mersenne, John Amos Comenius would go on to discuss universal language in *Via lucis* (not available in manuscript until c. 1642, and in print only in 1668).[25] In 1666 Leipzig, a nineteen-year-old Leibniz published *Dissertatio de arte combinatoria*, a short treatise that included a first sketch of a project for a philosophical language. Leibniz would continue to think and write about the subject for most of his life; in the fragment *Lingua generalis*, Leibniz argues that complex ideas can be transmitted through large numbers provided that simple concepts are previously assigned primary numbers. Leibniz abandoned this approach after 1678, and then sought alternatives taking Chinese characters as models.[26]

In Britain, Seth Ward wrote *Vindiciae academiarum* (1654), in which he proposed the construction of a universal language that represented the nature of things, and in 1661, while at Trinity College, Cambridge, the young Isaac Newton sketched a universal language and a phonetic alphabet, which he nonetheless never developed or completed.[27] Sir Thomas Urquhart, translator of Rabelais, also provided rules (yet no examples) for the formation of a universal language in *Ekskubalauron or The Discovery of a Most Exquisite Jewel* (1652) and *Logopandecteision, or An Introduction to the Universal Language* (1653).[28] In 1656 George Dalgarno began working on his proposal for a universal language. The outcome was *Ars signorum, vulgo character universalis et lingua philosophica* (1661), which organized concepts in seventeen classes and marked each class with a capital letter that all the concepts under the same class had as initial; each class was then subdivided into others that were represented by other letters. Ultimately, names for concepts were spelt by putting together all the letters of all the categories to which they belonged. However, John Wilkins's more impressive *An Essay towards a Real Character and a Philosophical Language* (1668), a work of 454 folio pages with long classificatory tables, soon came to outshine the *Ars signorum*.[29]

2. JOHN WILKINS'S *ESSAY* (1668) AND IDEAS ON LANGUAGE IN THE SEVENTEENTH AND EIGHTEENTH CENTURIES

John Wilkins (1614–1672) was a founding member of the Royal Society, presiding at four of its meetings (October 23 1661, May 21 and 28 and June 4, 1662) prior to the royal charter being sealed in July 1662, and he was also a member of five of the eight important committees organized in 1664 to direct the Society's activities (Mechanics, Anatomy, "Georgical" or Agricultural, History of Trade, and Correspondence).[30] In October 1662, the Royal Society assigned Wilkins the task of designing a universal language, and Wilkins duly set to work with the idea of devising a language that reflected reality as a mirror. Thus, not only would language be a means of communicating and acquiring knowledge, but it would also become knowledge itself: knowing a word would mean knowing the particularities of the thing it denoted. The universal "real character" would allow "the expression of our Conceptions by *Marks* which should signifie *things*, and not *words*".[31] Wilkins's philosophical language would be superior to any natural language because it would be based on things rather than words, and for Wilkins "things *are better then* words" and "real knowledge *is beyond* elegancy of speech".[32]

Wilkins's *Essay* is divided into four parts: the first identifies the defects and flaws of natural languages, allegedly absent from the philosophical language; the second classifies human concepts into forty categories; the third explains natural or universal grammar; and the fourth illustrates the application of the general rules.[33] William Lloyd's dictionary of symbols and words closes Wilkins's work. Additionally, the scientists Francis Willughby and John Ray collaborated with Wilkins on his project: they assembled the tables of insects, plants, and animals, and Ray rendered the *Essay* into Latin, a translation he began in December 1668 and completed by 1674, but which was never published and is now lost.[34]

In the end the Royal Society did not support the printing of the book, although it did lend Wilkins its imprimatur.[35] The fire of London of 1666 had destroyed most of the manuscript of the *Essay*, "and all but two copies of the portion of the work that had been printed".[36] Despite Wilkins's efforts, the *Essay* did not prove a success, not even within the circle of the Royal Society, which in May 1668 appointed a committee of linguists and scientists to assess the *Essay*. The committee included the phonetician William Holder, author of *The Elements of Speech* (1669); the mathematician John Wallis, who had written the grammar *Grammatica linguae anglicanae* (1653); Thomas Henshawe, co-author of the etymological dictionary *Etymologicon linguae anglicanae* (1671); the naturalist John Ray; the chemist Robert Boyle; the architect Christopher Wren; John Wilkins's half-brother, Walter Pope; the bishop of Salisbury and previous professor of astronomy at Oxford, Seth Ward; and the curator of experiments for the Royal Society, Robert Hooke. There is no report of the meetings of these men, and in fact no evidence they ever met. Their silence may have reflected their preference to avoid a negative report on Wilkins's project. Yet both Robert Hooke and John Ray seemed genuinely interested in it.[37] Along with Andrew Paschall, Francis Lodwick, Thomas Pigott (fellow of Wadham College, Oxford), and the antiquarian John Aubrey (fellow of the Royal Society), they agreed to work on Wilkins's proposal with a view to improving its classifications of objects and concepts, elaborating a dictionary-index, and devising more efficient symbols. Despite their good intentions, however, the group's projects eventually came to nothing.[38]

Of course, Wilkins's enterprise was flawed for a number of reasons. It did not acknowledge that ideologies and cultures inevitably determine the mental framework of speakers and that even the creator of an ideally faultless philosophical language would necessarily suffuse it with his particular cultural and ideological background.[39] Not to mention that "it is fundamentally impossible to draw up a complete and universally valid classification scheme encompassing the entire universe", one, moreover, that could be converted into classificatory tables easy to memorize and "accommodate to flexible needs of language users and to changes in our knowledge of the world".[40] Nevertheless, that Wilkins's project was impracticable does not mean the *Essay* was fruitless or uninfluential. Much to the contrary, the *Essay* contributed to the standardization of scientific nomenclature, had an impact on later grammarians such as Christopher Cooper, author of *Grammatica linguae anglicanae* (1685), and inspired Roget in the process of compiling his *Thesaurus* (1852), and its study of phonetics and the alphabet remained authoritative for many years after Wilkins's death.[41] Leibniz's interest in Wilkins's studies led him not only to write to the Royal Society, but to visit

London and discuss the matter with Boyle and Oldenburg.[42] Wilkins even played a key role in the establishment of the plain and simple style the Royal Society advocated for scientific language,[43] which obeyed their "philosophy of things, not of words".[44]

Abraham Cowley's poem "To the Royal Society", which preceded Thomas Sprat's *History of the Royal Society* (1667), encapsulates in three lines the suspicion of the Royal Society towards words and rhetoric and its antithetical trust in things: "From Words, which are but Pictures of the Thought, / (Though we our Thoughts from them perversly drew) / To Things, the Minds right Object".[45] Sprat himself acknowledges the existence of "a constant Resolution" on the part of the Royal Society "to reject all the amplifications, digressions, and swellings of style: to return back to the primitive purity, and shortness, when men deliver'd so many *things*, almost in an equal number of *words*".[46] To achieve this end, members of the Royal Society were recommended "a close, naked, natural way of speaking; positive expressions; clear senses; a native easiness: bringing all things as near the Mathematical plainness" as possible; "the language of Artizans, Countrymen, and Merchants, before that, of Wits, or Scholars" was at all times preferred.[47] The Royal Society was not alone in its predilection for plainness and simplicity in linguistic expression; this was a shared trend in seventeenth-century Europe, a reaction against the complicated and extravagant style and flowery expression of the late Renaissance which reached the British Isles circa 1660. The rise of scientific writing in the vernacular undoubtedly helped in establishing simple referential prose as the central style in Modern English.[48]

Seventeenth-century linguistic ideals revolved around the conviction that the only good language was rigid, static, and alien to change. Prescriptive linguistics prevailed over descriptive approaches, grammatical rules were thought essential, unchanging Latin and Greek became the models to follow, and the idea of creating an academy of the language, in emulation of the French, appealing. Proposals for an academy would finally come to nothing, but the seventeenth century prepared the ground for the avalanche of prescriptive grammars and comprehensive dictionaries that flourished during the eighteenth. Samuel Johnson's *Dictionary* (1755) became the jewel in the crown, with Johnson's initial project aiming to permanently standardize the language. However, after realizing that this goal was unattainable, due to the ever-changing nature of language, Johnson finally admitted that trying to preserve "words and phrases from mutability" and "secure" language "from corruption and decay" was as feasible as "to change sublunary nature, or clear the world at once from folly, vanity, and affectation".[49]

Forty-three years before Johnson's endeavour to fix the language, Jonathan Swift had written his *Proposal for Correcting, Improving, and Ascertaining the English Tongue* (1712), where he made manifest his disgust towards the corruption of English letters. According to Swift, it was evident when looking back at the history of English that the language of his time was "extremely imperfect",[50] caught in the midst of an accelerated process of degeneration which caused the frantic changes it was undergoing. To prevent further decline, Swift advocated refining it "to a certain Standard" and then finding a way "to fix it for ever";[51] after all, it "is better a language should not be wholly perfect, than that it should be perpetually changing".[52] If Swift supported linguistic simplicity and elevated such works as *The Book of Common Prayer* and King James's translation of the Bible to the level of linguistic models, he vehemently criticized modish affectations and the language of the court and the theatre.[53] The *Proposal* would not be Swift's only work on the matter of language: essays and pamphlets such as *Hints Towards an Essay on Conversation* (1713), *Letter to a Young Clergyman* (1720), *On the Corruption of the English Tongue* (1720), *Discourse to Prove the Antiquity of the English Tongue* (1729), and "Polite Conversation" (1738) followed. Furthermore, some of Swift's literary works, including *A Tale of a Tub* (1704) and *Gulliver's Travels* (1726), also display his thoughts on language. For instance, Lemuel Gulliver's "plain simple Style" is praised on the first page of *Gulliver's Travels*,[54] and a few pages later, we again read that his "style is very plain and simple".[55] Because Swift was of the opinion that "unruly and licentious society infects and corrupts language" and that "improper and incorrect language subverts and weakens society", his defence of plain style can also be seen as a political statement.[56]

3. JONATHAN SWIFT'S *GULLIVER'S TRAVELS* AGAINST THE ROYAL SOCIETY AND UNIVERSAL LANGUAGE PROJECTS

Jonathan Swift actively engaged in the discrediting of the Royal Society, which from the time of its constitution was accused of focusing on trifles and producing anecdotal knowledge through ludicrous experiments.[57] In *Gulliver's Travels* the Royal Society is satirically represented in the form of the delirious Academy of Lagado, which is described, like the Royal Society, as being sanctioned by the country's monarch and as having the aim of revising traditionally accepted scientific principles. As Gulliver reports, the founders of the Academy "fell into Schemes of putting all Arts, Sciences, Languages, and Mechanicks upon a new Foot", and "To this End, they procured a Royal Patent, for erecting an Academy of PROJECTORS in *Lagado*".[58] The absurdity and fruitlessness of the extravagant experiments carried out by the peculiar scholars of the Academy are indeed remarkable, including attempts to extract sunbeams from cucumbers, to transform human excrement back into food, and to build houses from the roof down. Similarly, in *Tale of a Tub* there are references to the construction of an enormous Academy subdivided into bizarre and ostensibly pointless schools ("School of *Looking-Glasses*", "School of *Swearing*", "School of *Salivation*", etc.).[59]

In *Gulliver's Travels*, the odd innovations of the Academy of Lagado have dreadful consequences for society and the economy, and leave the country in "miserably waste, the Houses in Ruins, and the People without Food or Cloaths".[60] Conversely, the communities that resist changing their traditional way of life, and refuse to adopt the recommendations of the Academy, escape ruin; the implication here is that sticking to tradition is safer and wiser than following wild new projects. In political terms, Swift's message is that relying on tradition (i.e., supporting the Tory party) is the sensible thing to do, while it is reckless to endorse the Whigs.[61]

Swift's satirical attack on the Royal Society, however, is just the tip of a more general critique of the new scientific, intellectual, and religious tendencies of his age. For example, his pessimistic views of human nature contrasted with the prevailing optimism of the times, as the eighteenth century began to question, among other beliefs, the limitations of human reason and the fallen state of mankind.[62] Swift stood by faith in divine revelation and the Scriptures, putting common sense before speculative philosophy and old forms of humanistic learning before the new experimental science. In short, he clung to the past and venerated it fiercely; he was sceptical about progress, especially for a corrupted and fallen being such as Man, and doubted that literature and art could have purposes other than moral ones.[63] In the dichotomy between tradition and modernity, Swift assuredly supported tradition.[64]

Swift's criticism of the Royal Society includes its linguistic projects. Gulliver notices an underlying confusion between words and things in virtually all the linguistic experiments carried out at the Academy of Lagado. The Academy's schemes to improve the language involved abridging discourse by shortening polysyllables to monosyllables, "and leaving out Verbs and Participles; because in Reality all things imaginable are but Nouns".[65] What is more, the Academy advocated the complete elimination of words because, given that "Words are only Names for *Things*, it would be more convenient for all Men to carry about them, such *Things* as were necessary to express the particular Business they are to discourse on".[66] That is, the Academy supported communication by means of things instead of words, which had the phenomenal advantage of it working "as an universal Language to be understood in all civilized Nations".[67] The Academy's scheme had but a single acknowledged "Inconvenience attending it":

> if a Man's Business be very great, and of various Kinds, he must be obliged in Proportion to carry a greater Bundle of *Things* upon his Back, unless he can afford one or two strong Servants to attend him. I have often beheld two of those Sages almost sinking under the Weight of their Packs, like Pedlars among us; who when they met in the Streets would

> lay down their Loads, open their Sacks, and hold Conversation for an Hour together; then put up their Implements, help each other to resume their Burthens, and take their Leave.[68]

This proposal, based on the use of physical objects rather than words for communication, reflects the desire of the Royal Society to draw words and things closer together so that, ideally, there would be an equal number of both. Gulliver also describes the mad invention of a specialist in "speculative learning" which is intended to accomplish "a complete Body of all Arts and Sciences" by means of a large, complex frame handled by forty of his pupils:

> [The frame] was Twenty Foot square, placed in the Middle of the Room. The Superficies was composed of several Bits of Wood, about the Bigness of a Dye, but some larger than others. They were all linked together by slender Wires. These Bits of Wood were covered on every Square with Paper pasted on them; and on these Papers were written all the Words of their language in their several Moods, Tenses and Declensions, but without any Order. The Professor then desired me to observe, for he was going to set his Engine at work. The Pupils at his Command took each of them hold of an Iron Handle, whereof there were Forty Fixed round the Edges of the Frame; and giving them a sudden Turn, the whole Disposition of the Words was entirely changed. He then commanded Six and Thirty of the Lads to read the several Lines softly as they appeared upon the Frame; and where they found three or four Words together that might make Part of a Sentence, they dictated to the four remaining Boys who were Scribes. This Work was repeated three or four Times, and at every Turn the Engine was so contrived, that the Words shifted into new Places, as the square Bits of Wood moved upside down.
>
> Six Hours a-Day the young Students were employed in this Labour; and the Professor showed me several Volumes in large Folio already collected, of broken Sentences, which he intended to piece together and out of those rich Materials to give the World a complete Body of all Arts and Sciences, which however might be still improved, and much expedited if the Publick would raise a Fund for making and employing five Hundred such Frames in *Lagado*, and oblige the Managers to contribute in common their several Collections.[69]

By means of this episode, Swift also criticizes the faith which the Royal Society placed in mechanical operations contrary to common sense, and the thought that, in the process of finding answers to the most fundamental questions, human intelligence can be replaced by an unpredictable mechanical apparatus. One of the underlying flaws of this invention is that it mistakes words for things in that the random connections between words established by the a-logical device are understood as actual relations between the things for which they stand. For Swift, the confusion between words and things doomed the project of a universal language too. By this time John Locke had emphasised the distinction between words and things and pointed out the arbitrary nature of language, an observation which would not only account for linguistic diversity but for the fact that no natural universal language existed:

> *Words*, which were by nature so well adapted to that purpose, came to be made use of by men as the signs of their ideas; not by any natural connexion that there is between particular articulate sounds and certain ideas, for then there would be but one language amongst all men; but by a voluntary imposition, whereby such a word is made arbitrarily the mark of such an idea. The use, then, of words, is to be sensible marks of ideas; and the ideas they stand for are their proper and immediate signification.[70]

By attaching names to ideas, Locke continues, humans "consider things, and discourse of them, as it were in bundles, for the easier and readier improvement and communication of their knowledge, which would advance but slowly were words and thoughts confined only to particulars".[71] The controversy over words and things in the seventeenth and eighteenth centuries was certainly a major debate at the root of the "scientific revolution", if this revolution is understood as a "turning away from the world of 'words' towards that of 'things'".[72] Despite the numerous and profound differences between Swift and the members of the Royal Society (Wilkins included), and their opposing ideas regarding the new empirical science, they all share a marked preference for a plain and simple style and give pre-eminence to things over words. Swift's sympathies were also in fact on the side of things, in accordance with the general suspicion of words which was characteristic of the Augustan period. The reasoning here was that words could not be trusted because they were mutable, ever changing, abstract enough to deceive men, and in themselves a human invention (and as such, subject to human vices and corruptions and transmitters thereof). By contrast, things had a divine nature, for they came from God; from them, as the Royal Society argued, it was possible to obtain empirical knowledge.

4. CONCLUSION

While Wilkins was one of the founding members of the Royal Society, Swift became a merciless critic of it. Projects for a universal language are satirized by Swift, who highlights a structural confusion between words and things in such proposals. Nonetheless, both he and Wilkins agreed on the idea that language variation was a sign of linguistic corruption and decay. Wilkins warns that "if any *English* man should now write or speak as our forefathers did about six or seven hundred years past, we should as little understand him as if he were a foreigner"[73] because "this last Century may be conjectured to have made a greater change in our Tongue, then any of the former, as to the addition of new words".[74] Consequently, "if in some few hundreds of years a Language may be so *changed* as to be scarce intelligible", Wilkins speculates, "in a much longer tract of time it may be quite *abolished*".[75] For this reason, he concludes that "every *change* is a *gradual corruption*".[76]

For both Swift and Wilkins, then, linguistic mutability is undesirable, and their linguistic proposals operate, albeit differently, towards the same goal: controlling linguistic change and, if possible, fixing the language. The universal and philosophical language of Wilkins was intended to overcome the particular differences of natural tongues, which vary because of their arbitrary nature and the gap that separates them from things. The envisaged perfection of the philosophical language guarantees that, once a thing is represented by its character, as long as the nature of the thing remains the same, its assigned real character will not vary either. Similarly, Swift's desire to see the language fixed is consistent with his belief that linguistic change is, by definition, negative, as it removes language from its past and its history. The solution lies, in Swift's eyes, in preventing the language that already exists from further decay, instead of engaging in linguistic experiments of a universal nature, whose promoters would only employ it to disseminate and expand the new experimental science which Swift so adamantly criticized.

Appendix I

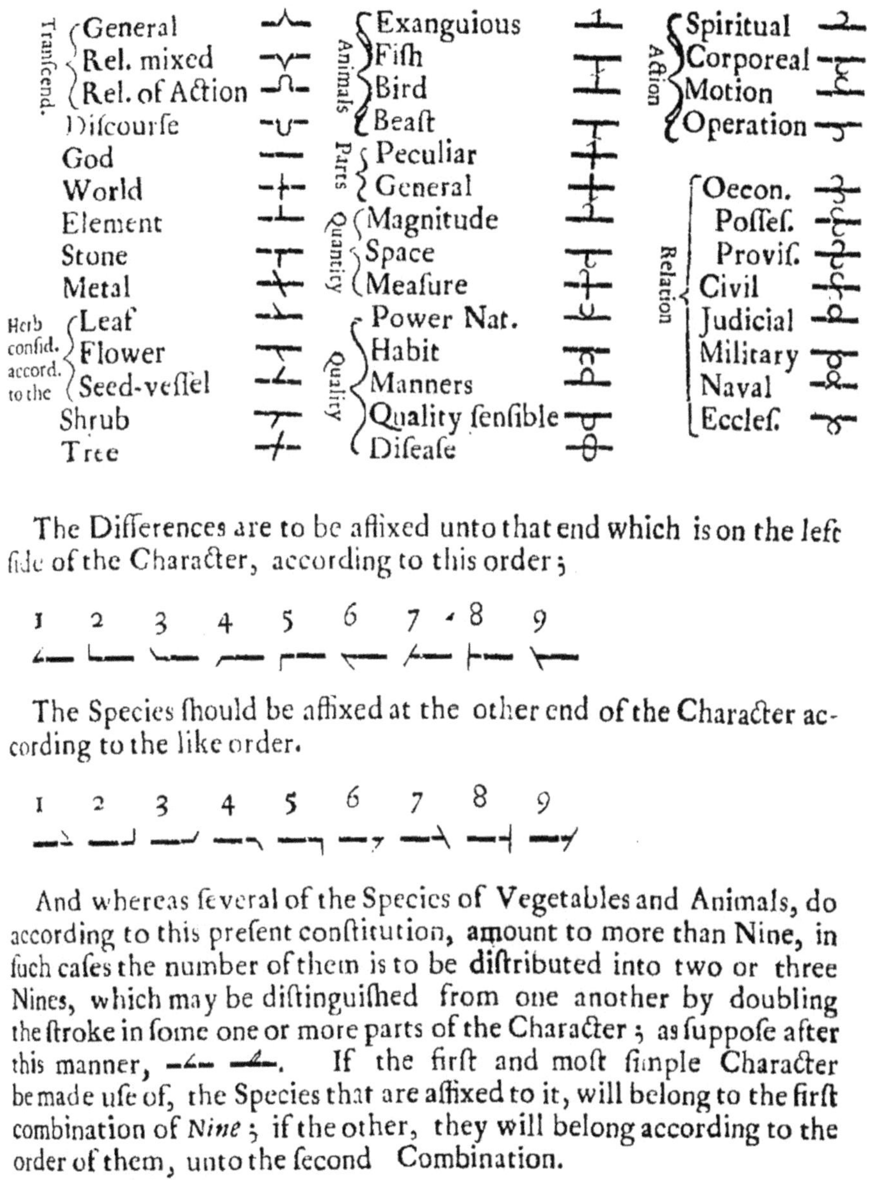

Tranſcend. General; Rel. mixed; Rel. of Action
Diſcourſe
God
World
Element
Stone
Metal
Herb confid. accord. to the Leaf; Flower; Seed-veſſel
Shrub
Tree

Animals Exanguious; Fiſh; Bird; Beaſt
Parts Peculiar; General
Quantity Magnitude; Space; Meaſure
Quality Power Nat.; Habit; Manners; Quality ſenſible; Diſeaſe

Action Spiritual; Corporeal; Motion; Operation
Relation Oecon.; Poſſeſ.; Proviſ.; Civil; Judicial; Military; Naval; Eccleſ.

The Differences are to be affixed unto that end which is on the left ſide of the Character, according to this order;

1 2 3 4 5 6 7 8 9

The Species ſhould be affixed at the other end of the Character according to the like order.

1 2 3 4 5 6 7 8 9

And whereas ſeveral of the Species of Vegetables and Animals, do according to this preſent conſtitution, amount to more than Nine, in ſuch caſes the number of them is to be diſtributed into two or three Nines, which may be diſtinguiſhed from one another by doubling the ſtroke in ſome one or more parts of the Character; as ſuppoſe after this manner, . If the firſt and moſt ſimple Character be made uſe of, the Species that are affixed to it, will belong to the firſt combination of *Nine*; if the other, they will belong according to the order of them, unto the ſecond Combination.

Image 11.1 Wilkins, *An Essay*, Ddd2r

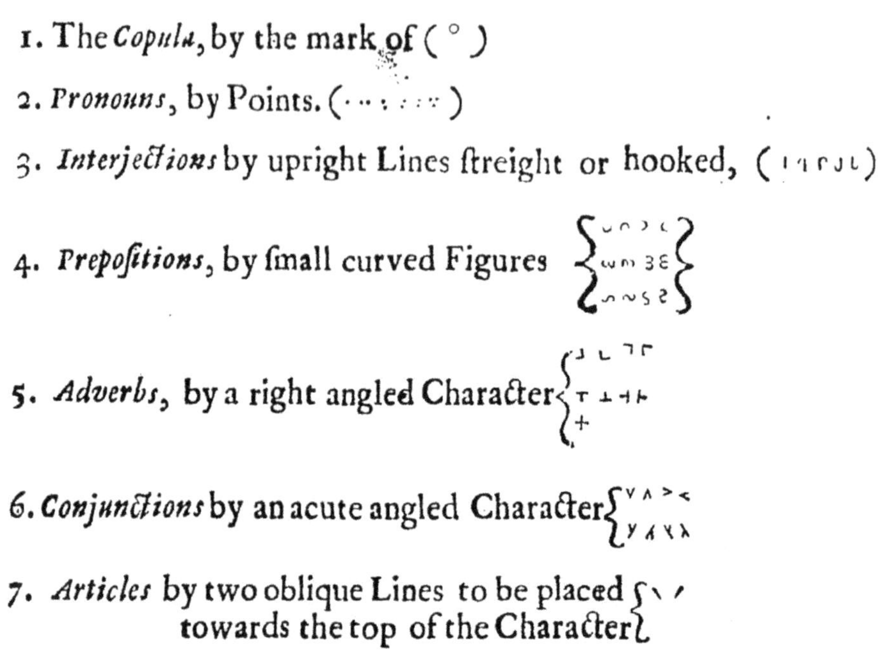

1. The *Copula*, by the mark of (°)

2. *Pronouns*, by Points.

3. *Interjections* by upright Lines ſtreight or hooked,

4. *Prepoſitions*, by ſmall curved Figures

5. *Adverbs*, by a right angled Character

6. *Conjunctions* by an acute angled Character

7. *Articles* by two oblique Lines to be placed towards the top of the Character

Image 11.2 Wilkins, *An Essay*, Ddd2[v]

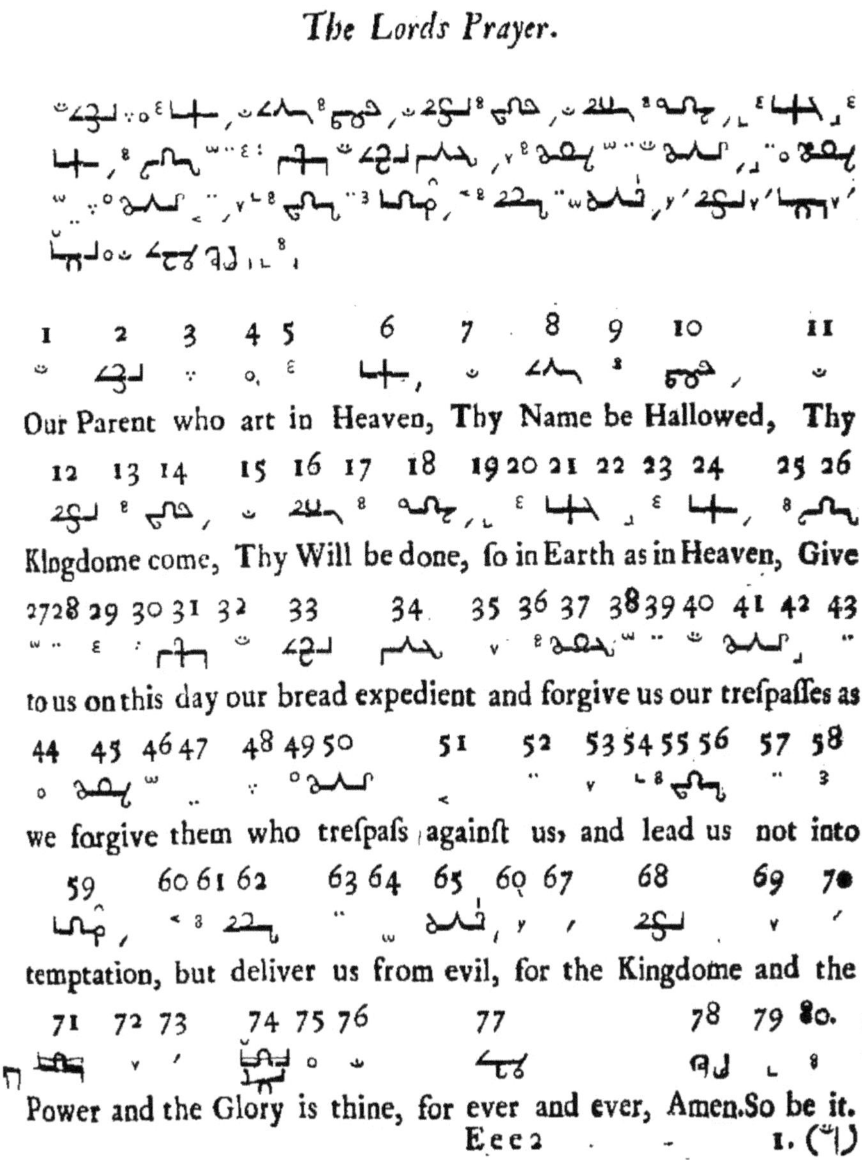

The Lords Prayer.

1 2 3 4 5 6 7 8 9 10 11
Our Parent who art in Heaven, Thy Name be Hallowed, Thy
12 13 14 15 16 17 18 19 20 21 22 23 24 25 26
Kingdome come, Thy Will be done, ſo in Earth as in Heaven, Give
27 28 29 30 31 32 33 34 35 36 37 38 39 40 41 42 43
to us on this day our bread expedient and forgive us our treſpaſſes as
44 45 46 47 48 49 50 51 52 53 54 55 56 57 58
we forgive them who treſpaſs againſt us, and lead us not into
59 60 61 62 63 64 65 66 67 68 69 70
temptation, but deliver us from evil, for the Kingdome and the
71 72 73 74 75 76 77 78 79 80.
Power and the Glory is thine, for ever and ever, Amen. So be it.

Eee2 1. (

Image 11.3 Wilkins, *An Essay*, Eee2r

NOTES

1. Peter Dear, "Totius in Verba: Rhetoric and Authority in the Early Royal Society," *Isis* 76, no. 2 (1985): 144–161; Diana M. Judd, *Questioning Authority: Political Resistance and the Ethic of Natural Science* (New Brunswick: Transaction Publishers, 2009).
2. Henry Knight Miller, "Henry Fielding's Satire on the Royal Society," *Studies in Philology* 57, no. 1 (1960): 72–86.
3. Frederick N. Smith, "Scientific Discourse: *Gulliver's Travels* and *The Philosophical Transactions*," in *The Genres of Gulliver's Travels*, ed. Frederick N. Smith (Newark: University of Delaware Press, 1990), 139.
4. James Knowlson, *Universal Language Schemes in England and France, 1600–1800* (Toronto and Buffalo: University of Toronto Press, 1975), 30–36.
5. Lynda Mugglestone, *The Oxford History of English* (Oxford: Oxford University Press, 2006), 237.
6. Jaap Maat, *Philosophical Languages in the Seventeenth Century: Dalgarno, Wilkins, Leibniz* (Dordrecht and London: Kluwer Academic, 2004), 21.
7. Knowlson, 16.
8. Vivian Salmon, *The Works of Francis Lodwick: A Study of His Writings in the Intellectual Context of the Seventeenth Century* (London: Longman, 1972), 129.
9. Quoted in Richard F. Jones, *The Seventeenth Century: Studies in the History of English Thought and Literature from Bacon to Pope* (Stanford: Stanford University Press, 1951), 152.
10. Athanasius Kircher elaborately discussed Egyptian hieroglyphs in *Lingua aegyptiaca restitute* (1643) and *Oedipus aegyptiacus* (1652–1655). Until the discovery of the Rosetta Stone in 1799, hieroglyphs remained undeciphered.
11. David Cram and Jaap Maat, eds. *George Dalgarno on Universal Language: The Art of Signs (1661), The Deaf and Dumb Man's Tutor (1680), and the Unpublished Papers* (Oxford: Oxford University Press, 2001), 4.
12. Sir Francis Bacon, *Collected Works of Francis Bacon, Vol. VII*, eds. James Spedding, Robert Leslie Ellis and Douglas Denon Heath (London: Routledge, 1996), 439.
13. Richard Foster Jones, "Science and Language in England of the Mid-Seventeenth Century," *Journal of English and Germanic Philology* 31, no. 3 (1932): 328.
14. Vivian Salmon, "The Evolution of Dalgarno's 'Ars Signorum'," in *Studies in Language and Literature in Honour of Margaret Schlauch*, eds. Mieczyslaw Brahmer, Stanislaw Helsztyński and Julian Krzyżanowski (Warsaw: Polish Scientific Publishers, 1966), 353.
15. John Wilkins, *Mercury, or, The Secret and Swift Messenger*, ed. Brigitte Asbach-Schnitker (Amsterdam; Philadelphia: John Benjamins, 1984), 89.
16. Wilkins, *Mercury*, 89.
17. For more on this work (and Dalgarno's *Ars signorum*), see Cram and Maat, eds.
18. Francis Lodwick, *On Language, Theology, and Utopia*, eds. Felicity Henderson and William Poole (Oxford: Clarendon Press, 2011).
19. Cave Beck listed words alphabetically and attached a number to each. The result was an international code which allowed translation simply by substituting, with the help of a dictionary, the number for the word in the target language.
20. The only extant fragments of this work are those included in Schott's *Technica curiosa*. For more on the contribution of Spanish authors to universal language schemes, see María Luisa Calero Vaquera, *Proyectos de Lengua Universal: La*

Contribución Espan~ola (Córdoba: Publicaciones de la Universidad de Córdoba y Obra Social y Cultural CajaSur, 1999).

21. For more on projects for a universal character in the seventeenth century, see Jonathan Cohen, "On the Project of a Universal Character," *Mind* 63, no. 249 (1954): 49–63. See also Knowlson and, of course, Rhodri Lewis, *Language, Mind and Nature: Artificial Languages in England from Bacon to Locke* (Cambridge: Cambridge University Press, 2007).
22. Philosophical languages necessarily had to be of the *a priori* type, "composed entirely of invented elements which are not found in any ancient or modern language", hence in contrast to artificial languages such as Esperanto or Interlingua, which are *a posteriori* in nature, that is, "based upon ele-ments of grammar, vocabulary and syntax drawn from one or more natural languages". J. Andrew Large, "A Real Character: Seventeenth-Century Universal Language Schemes," *Multilingua: Journal of Cross-Cultural and Interlanguage Communication* 2, no. 1 (1983): 6.
23. Joseph L. Subbiondo, "Educational Reform in Seventeenth-Century England and John Wilkins' Philosophical Language," *Language & Communication* 21, no. 3 (2001): 273.
24. Cram and Maat, *The Deaf and Dumb*, p. 6.
25. Benjamin DeMott, "Comenius and the Real Character in England," *PMLA* 70, no. 5 (1955), and "The Sources and Development of John Wilkins's Philosophical Language," *Journal of English and Germanic Philology* 57 (1958): 1–13.
26. Olga Pombo, *Leibniz and the Problem of a Universal Language* (Münster: Nodus Publikationen, 1987).
27. Ralph W. V. Elliott, "Isaac Newton's 'Of an Universall Language'," *The Modern Language Review* 52, no. 1 (1957): 1–18.
28. Urquhart claimed that the main body of his work had been lost in 1651 during the Battle of Worcester. Kelsie B. Harder, "Sir Thomas Urquharts Universal Language," *Notes and Queries* 3 (1956): 473–6.
29. A book totalling 636 folio pages—plus four fold-out plates—if we include 20 pages of front matter and 162 of an appended *Dictionary*. Interestingly, Dalgarno and Wilkins had worked together, but disagreed on the method to follow, and eventually broke off their collaboration. On Dalgarno's work, see Salmon, "The evolution"; Wayne Shumaker, *Renaissance Curiosa: John Dee's Conversations with Angels, Girolamo Cardano's Horoscope of Christ, Johannes Trithemius and Cryptography, George Dalgarno's Universal Language* (Binghamton, N.Y.: Center for Medieval and Early Renaissance Studies, 1982); and Cram and Maat, *George Dalgarno on Universal Language*, 27–29. For more on the connection between Dalgarno and Wilkins, see Cram, "George Dalgarno on 'Ars Signorum' and Wilkins' 'Essay'," in *Progress in Linguistic Historiography: Papers from the International Conference on the History of the Language Sciences (Ottawa, 28–31 August 1978)*, ed. Konrad Koerner (Amsterdam: John Benjamins, 1980), 113–121, and Joseph L. Subbiondo, "Competing Models for a 17th Century Universal Language: A Study of the Dispute between George Dalgarno and John Wilkins," in *History of Linguistics 2005*, ed. Douglas A. Kibbee (Amsterdam: John Benjamins, 2007), 113. Subbiondo particularly discusses the "breach of judgment" between Dalgarno and Wilkins that came to a head in the first months of 1658.
30. Dorothy Stimson, "Dr. Wilkins and the Royal Society," *The Journal of Modern History* 3, no. 4 (1931): 550–552.
31. John Wilkins, *An Essay towards a Real Character, and a Philosophical Language* (London: Printed for Sa. Gellibrand and for John Martyn, 1668), D3r. Italics in quoted extracts from the *Essay* are Wilkins's.
32. Wilkins, *An Essay*, a2r.

33. Appendix I contains three images taken from Wilkins's proposal for a philosophical language. The first shows the type of signs that mark different categories or properties of objects (such as materials); the second, grammatical categories of words. The third image captures a translation of the Lord's Prayer from English into Wilkins's philosophical language.
34. Otto Funke, "On the Sources of John Wilkins's *Philosophical Language* (1668)," *English Studies* 40, no. 1 (1959): 208–214; Mary Slaughter, *Universal Languages and Scientific Taxonomy in the Seventeenth Century* (Cambridge: Cambridge University Press, 1982), p. 176.
35. Rhodri Lewis, "The Publication of John Wilkins's *Essay* (1668): Some Contextual Considerations," *Notes and Records of the Royal Society* 56 (2002): 133–146.
36. Barbara J. Shapiro, *John Wilkins, 1614–1672; An Intellectual Biography* (Berkeley: University of California Press, 1969), 213.
37. Vivian Salmon, "John Wilkins' *Essay* (1668). Critics and Continuators," in *John Wilkins and 17th-Century British Linguistics*, ed. Joseph L. Subbiondo (Amsterdam; Philadelphia: John Benjamins, 1992), 349–364.
38. Sidonie Clauss, "John Wilkins' Essay toward a Real Character: Its Place in the Seventeenth-Century Episteme," *Journal of the History of Ideas* 43, no. 4 (1982): 531–553; Rhodri Lewis, "The Efforts of the Aubrey Correspondence Group to Revise John Wilkins's *Essay* (1668) and their Context," *Historiographia Linguistica* 28 (2001): 333–366.
39. Sandrine Sorlin, "La Langue Philosophique de John Wilkins (1614–1672): Langage Universel ou Utopie Linguistique?," *Etudes Epistémè* 12 (2007): 118.
40. Maat, 142.
41. Clark Emery, "John Wilkins' Universal Language," *Isis* 38 (1948): 184–5; Shapiro, 220–222; Fredric Dolezal, *Forgotten but Important Lexicographers: John Wilkins and William Lloyd. A Modern Approach to Lexicography before Johnson* (Tübingen: Niemeyer, 1985).
42. Philip Beeley, "A Philosophical Apprenticeship: Leibniz's Correspondence with the Secretary of the Royal Society, Henry Oldenburg" in *Leibniz and his Correspondents*, ed. Paul Lodge (Cambridge: Cambridge University Press, 2004), 47–73.
43. Shapiro, 206; Francis Christensen, "John Wilkins and the Royal Society's Reform of Prose Style," in *John Wilkins and 17th-Century British Linguistics*, ed. Subbiondo, 133–154.
44. The Royal Society's "aim was to produce results beneficial to men, not to persuade them to accept one set of notions in place of another. Accordingly, in presenting the results of their investigations they sought, whether in speaking before the Society or in writing for their registers, simply to explain and record. The fundamental principle underlying their idea of style was that their results must stand or fall by their own evidence". Christensen, 146. For more on the English prose style advocated by the Royal Society, see Richard F. Jones, "Science and English Prose Style in the Third Quarter of the Seventeenth Century," *PMLA* 45, no. 4 (1930): 977–1009, and Brian Vickers, "The Royal Society and English Prose Style: a Reassessment," in *Rhetoric and the Pursuit of Truth: Language Change in the Seventeenth and Eighteenth Centuries*, eds. Brian Vickers and Nancy S. Struever (Los Angeles: William Andrews Clark Memorial Library, University of California, 1985), 3–76. See also the recently published Tina Skouen and Ryan Stark, eds. *Rhetoric and the Early Royal Society: A Sourcebook* (Leiden; Boston: Brill, 2014). Except the introduction, all the articles in this volume have been published previously.
45. Thomas Sprat, *The History of the Royal-Society of London for the Improving of Natural Knowledge* (London: Printed by T. R. [Thomas Roycroft], 1667), B2[r].

46. Sprat, P1r.
47. Sprat, P1r.
48. Albert C. Baugh and Thomas Cable, *A History of the English Language* (London: Routledge, 2003), 253–295.
49. Samuel Johnson, *Johnson's Dictionary: A Modern Selection*, eds. E. L. McAdam, Jr., and George Milne (Mineola: Dover Publications, 2005), 24.
50. Jonathan Swift, "A Proposal for Correcting, Improving, and Ascertaining the English Tongue", in *The Prose Works of Jonathan Swift. Vol. XI. Literary essays*, ed. Temple Scott (London: George Bell and Sons, 1907), 6.
51. Swift, *A Proposal*, 9.
52. Swift, *A Proposal*, 15.
53. Herbert John Davis, *The Satire of Jonathan Swift* (Westport, CT: Greenwood Press, 1979), 17–18.
54. Jonathan Swift, *Gulliver's Travels*, ed. David Womersley (New York: Cambridge University Press, 2012), 5.
55. Swift, *Gulliver's Travels*, 15.
56. Denis Donoghue, *Jonathan Swift: A Critical Introduction* (Cambridge: Cambridge University Press, 1969), 133. Deborah Baker Wyrick, *Jonathan Swift and the Vested Word* (Chapel Hill: University of North Carolina Press, 1988), 42. While from Swift's perspective a good language is indicative of a virtuous community of speakers, a language in permanent flux is instead symptomatic of vice and immorality, Rocío G. Sumillera, "Correlations between Language and Morality in *Gulliver's Travels*," in *Metamorphoses of Travel Writing: Across Theories, Genres, Centuries and Literary Traditions*, eds. Grzegorz Moroz and Jolanta Sztachelska (Newcastle: Cambridge Scholars Publishing, 2010), 146–155. For more on languages in seventeenth-century imaginary voyages, see Edward D. Seeber, "Ideal Languages in the French and English Imaginary Voyage," *PMLA* 60, no. 2 (1945): 586–597; and Paul E. Cornelius, *Languages in Seventeenth and Early Eighteenth-Century Imaginary Voyages* (Gene`ve: Librairie Droz, 1965).
57. Miller, 81; Marjorie Hope Nicolson and Nora M. Mohler, "The Scientific Background of Swift's Voyage to Laputa," *Annals of Science* 2 (1937): 299–334; Colin Kiernan, "Swift and Science," *The Historical Journal* 14, no. 4 (1971): 709–722; David Renaker, "Swift's Laputians as a Caricature of the Cartesians," *PMLA* 94, no. 5 (1979): 937.
58. Swift, *Gulliver's Travels*, 255–256.
59. Jonathan Swift, *A Tale of a Tub; The Battle of the Books; The Mechanical Operation of the Spirit*, ed. Frank H. Ellis (Frankfurt am Main: Peter Lang, 2006), 13.
60. Swift, *Gulliver's Travels*, 256.
61. Arthur E. Case, *Four Essays on Gulliver's Travels* (Gloucester, MA: P. Smith, 1958), 88–89.
62. Roland Mushat Frye, "Swift's Yahoos and the Christian Symbols for Sin," *Journal of the History of Ideas* 15 (1954): 201–217; Theodore O. Wedel, "On the Philosophical Background of *Gulliver's Travels*," in *Swift: Gulliver's Travels: A Casebook*, ed. Richard Gravil (London: Macmillan, 1974), 83–99.
63. Brian Tippett, *Gulliver's Travels* (Basingstoke: Macmillan, 1989), 44–45.
64. The well-known controversy about the ancients and moderns also appears in other works by Swift: *The Battle of the Books* describes in a mock heroic fashion a battle between the ancient and the modern volumes of St. James's library, and the disorganized and fragmented discourse of the narrative persona of *A Tale of a Tub* discredits his disdainful attitude towards the achievements of the ancients and his constant praise of modernity. Joseph M. Levine, "Stroke and Counterstroke", *The Battle of the Books: History and Literature in the Augustan Age* (Ithaca: Cornell University Press, 1991), 85–120.

65. Swift, *Gulliver's Travels*, 270.
66. Swift, *Gulliver's Travels*, 271.
67. Swift, *Gulliver's Travels*, 273.
68. Swift, *Gulliver's Travels*, 272.
69. Swift, *Gulliver's Travels*, 268–269.
70. John Locke, *An Essay Concerning Human Understanding, Vol. II*, ed. Alexander Campbell Fraser (New York: Dover Publications, 1959), 8–9 (III, 2, 1).
71. Locke, 31 (III, 3, 20).
72. Adrian Johns, "Reading and Experiment in the Early Royal Society," in *Reading, Society and Politics in Early Modern England*, eds. Kevin Sharpe and Steven Zwicker (Cambridge: Cambridge University Press, 2003), 244.
73. Wilkins, *An Essay*, B3^{v}.
74. Wilkins, *An Essay*, B4^{v}.
75. Wilkins, *An Essay*, B4^{v}.
76. Wilkins, *An Essay*, B4^{v}.

12 One Second Language for Mankind

The Rise and Decline of the World Auxiliary Language Movement in the Belle Époque

Markus Krajewski

The so-called scientific revolution of the seventeenth century was—like every new paradigm in the history of scientific thought—based on a new language. Scholars like Descartes and Leibniz referred predominantly to mathematics when abstracting their geometrical observations, and hence mathematics became their "universal" language for the description of problems in the physical or natural realm. They also made use of another language when they generalized their own systems of knowledge; Leibniz, Descartes, and John Wilkins were among those who developed artificial language schemes for scientific thought. While for Descartes the question had already arisen as to how such a universal language could map "pure" ideas onto "true" philosophy, the same question was discussed during the heyday of the universal language boom with a more formalized *calculus* which allowed for the almost mechanical proposing of true statements based on a specific semiotic system. This quest, of course, was pursued in vain, though Leibniz and Wilkins were fully convinced that they were very close to devising an effective universal language. The reason for the loss of interest in universal languages during the eighteenth century was at least twofold. On the one hand, one of the crucial promises of universal language schemes proved to be false: universal languages were not easy to learn. On the contrary, they were rather complicated due to their abstract nature and their design for mainly philosophical purposes. On the other hand, the scientific and scholarly use of national languages such as English, German, and French expanded in this period. Moreover, under the absolute reign of Louis XIV, French became the international language of ambassadors and the court, the new *lingua franca* amongst the well-trained upper class. In the age of imperial powers, and at the dawn of colonialism, a neutral, politically independent medium for universal communication and philosophical thought no longer seemed to be of any real need.

The history of universal languages for scientific purposes can be understood as the effect of a double profanation. First, at the risk of being accused of heresy, scholars such as Descartes and Comenius were successful in recalling the problem of universal understanding from the responsibility of God, moving it from its former theological realm towards an epistemic one. Second, the philosophical quest for universal communication gradually lost its metaphysical status when utopian plans proved impossible to be turned into practice. General language schemes gathered momentum once more in the second half of the nineteenth century, when the old problem of a universal language was again pursued, but under drastically different conditions and independent from the closed circles of academies and learned societies. This time a discourse was built not in the name of a *universal* language, but was presented in a more modest and less inscrutable fashion: by using the programmatic term *world auxiliary language*, a globally standardized language project initiated around 1900 by the so-called projectors (or "project makers").[1] This global network developed during the second half of the nineteenth century in parallel to the postal and telegraph networks, the steam ship lines, the railway system, and other forms of transport and communication.

In this early process of globalization, Wilhelm Ostwald played a major role. Winner of the 1909 Nobel Prize for Chemistry, he was one of the first official German visiting professors at Harvard, where he was invited by Hugo Münsterberg to spend a year researching along with American colleagues. In 1904, during his stay, Ostwald visited the World Exposition at St. Louis, afterwards noting that it produced in him "a deep and lively global feeling of being at home, when one comes together with men on such occasions for the first time whose achievements are known from literature and, after the initial greetings, at once finds himself on common, mutually familiar ground".[2] However, despite this euphoric "global feeling" of being at home, Ostwald identified a significant and recurrent problem in lectures, plenary discussions, and in conversations:

> There stood next to one another the men who had the most important things to say to each other, but they could not come to an understanding. For even if most scholars and practitioners today master multiple languages sufficiently to be able to read technical papers, it is still a long and arduous journey from this point to oral communication in the foreign language. Thus, from this necessity arose anew the idea of international languages.[3]

In order to remedy the glaring drawback of mastering oral expression, the mathematician and philosopher Louis Couturat, of the Collège de France, together with the later co-author of his history of universal language, Léopold Leau, convened a commission. At the 1900 World Fair they founded the *Délégation pour l'adoption d'une langue auxiliaire internationale* with the goal of choosing the best of the existing global auxiliary languages and subsequently presenting it to the entire world as the sole alternative to natural languages as the international means of communication. The choice was anything but easy. At the turn of the twentieth century there existed approximately 250 so-called "planned" or "artificial languages". The Délégation also made it their task to win renowned scholars and scientists over to their side. Thus, a message from Couturat, who became aware of Ostwald's interest in artificial languages, also made its way to the University of Leipzig. At the time, Wilhelm Ostwald's view was that the learning of multiple foreign languages was an example of a waste of energy, one which might easily be spared by accepting a unified auxiliary language. Indeed, the Riga-born Ostwald saw himself as the victim of a childhood linguistic confusion, as he spoke German as a first language, Russian in school, and Latvian in day-to-day life. Moreover, through one of his teachers at the University of Dorpat, Estonia, he became acquainted with the artificial language Volapük. Thus Ostwald eagerly pledged his cooperation with the Délégation and would deploy unrelenting propaganda for the idea of a global auxiliary language in the years prior to the First World War. His advertising activities ranged from lectures to the Bavarian regional assembly of the Society of German Engineers to the establishment of nearly 100 American Esperantist-Clubs, founded in the wake of Ostwald's semester abroad at Harvard from 1905–1906.

Central to the aims of many of the projectors, and despite the often heterogeneous nature of how the task was itself understood, was the idea of advancing the "welfare of entire empires, indeed of the whole world" through a universal language, as Johann Gottlieb von Justi expressed it.[4] As soon as the projector made it his goal to give his plans a broad validity extending beyond the regional sphere, a language with certain characteristics became indispensable; to wit, it needed to be easily learned and understood and identical throughout the world.

The preoccupation with a general language in the second half of the nineteenth century did not fall within the discourse of "universal language", but rather adopted the programmatic title of *world auxiliary language*. From this point on, and in parallel to the expansion of the transport and communications infrastructure, the aim of this movement consisted of providing a universal means of communication. In contrast to the strategies of Gottfried Wilhelm Leibniz or John Wilkins in the seventeenth century, the nineteenth-century projectors attempted neither to find a highly sophisticated philosophical instrument of insight nor to reconstruct a proto-language. Rather, a new code was to be developed wholly to serve the purposes of communication in the realms of trade, industry, and science. Louis Couturat, the philosopher, mathematician, and trailblazer of the world auxiliary language movement, noted in *Histoire de la langue universelle* (1903) that the expansion of national markets into international ones, as well as technological advances, was opposed by the simultaneous atrophy of the "mental" means of communication. Thus what was needed was a kind of "mental bond" to keep pace with the technical and practical means available in the world, which, by the turn of the twentieth century, had long been superior: "What good is it if we can travel, write, and speak from one end of the world to the other if we cannot understand each other?"[5]

Thus the world auxiliary language scheme not only aimed at becoming known world-wide, but also to establish itself as the only universal second language. Indeed, according to Couturat, the international auxiliary language should be "the second language for every man".[6] Couturat identified three target groups which would primarily profit from the advantages of such a code: scholars, merchants, and travelers (the distinction between the last two is difficult to determine at times). However, the general goal was to establish this language among all spheres of society so that it did not merely become "a technical or aristocratic language only accessible to a few initiates", but rather "a daily language which finds use on the railway and in the hotel just as in the learned societies and at conferences".[7]

It would have been logical for Couturat, the editor of the works of Leibniz, to align himself with an existing language; yet choosing a natural language such as English or French would not have been politically wise in that neutrality would not have been preserved and one specific linguistic community would have had a clear and undue advantage over the rest. In the words of Wilhelm Ostwald: "The selection of any natural language is precluded for the simple reason that they are all no good".[8] With the simplification or adaptation of classical languages such as Latin, the situation would have been similar: Latin could certainly be spoken by scholars, but for "every man" it would have been too difficult, given its numerous exceptions, special rules, and "superfluous complications". Plus, the logical introduction of new technical terms would present considerable difficulties.[9] Interestingly, Friedrich Nietzsche also wondered about a supposedly remote linguistic scenario:

> And in some distant future there will be a new language, first as a commercial language, then as a language of intellectual communication overall, for everyone, as surely as there was once airship travel. Why would linguistics have studied the laws of language for a century and assessed that which is necessary, is valuable and succeeds in every individual language?[10]

Nietzsche could not have known that the very same year, such a distant future had already begun. At least, it had for a village priest in Baden, southern Germany, who sought a language specifically designed for the purposes of global communication, one which was easy to learn, simple to speak, and able to satisfy the relentless lexical growth of the times by easily allowing compound formation.

Couturat and Ostwald were aware of models for a global auxiliary language to be constructed, and indeed those actually established in the international code of symbols common since 1857, as well as in algebraic and chemical formulas and musical notation. In addition to the unambiguousness of these artificial communication codes, as in Samuel Morse's telegraph alphabet, the advocates of artificial languages praised the high degree of structural regularity, the absence of exceptions, as well as the lack of redundancy in such systems of symbols. All manner of superfluous information (such as plurals and gender inflection) can be avoided in a tailored language. Of course, the difference between systems such as Morse Code, the international symbol code, and the railroad signal system on the one hand, and natural languages on the other, is that the former are pasigraphies, that is, systems of signals—a difference that, interestingly, goes largely unnoticed by the advocates of a world auxiliary language. Ultimately it is a matter of proving the extent to which communication codes can be spread across the entire world and of the program of their implementation. Even if the pamphlets in favor of a world auxiliary language belong to a period in which there was an abundance of projected plans, with little stress on "feasibility studies" by specialists, their authors knew that evidence had to be presented to strengthen the authority of a global language and hence the necessity of its dissemination.

Within the scientific community, Carl Friedrich Gauss created an unprecedented structure for the successive propagation of newly discovered physical quantities with his Magnetic Society, supposedly the first international consolidation of scientific aims. This model was then adopted by the "Assembly for the Determination of International Electrical Units" in Paris[11] in order to make it accessible through linking it to technical practice and "ultimately with national authority for general use".[12] This shows how the power of definition lies with those who first recognize the requirements and take the initiative, leading eventually to general recognition through national governments.

In the development of world auxiliary languages in the late nineteenth century, the foundation of the Postal Union in Bern in 1874 became the most significant institution in terms of the global standardization for internationalism. The Baden prelate Johann Martin Schleyer (1831–1912) developed his so-called universal alphabet, which was intended to combine global communication brought about by the global postal service with the most varied systems of notation. Though this alphabet was temporarily denied its desired effect, in 1879 Schleyer, from his parsonage in Litzelstetten on Lake Constance, dreamed of the continued development of the notation into nothing less than a global language, which he devised on the basis of a corrupted version of English. He called his new construct Volapük, from *vol* (world) and *pük* (speak), that is, world language. Its aim was that of "helping the global postal system, this magnificent accomplishment of the modern era, to achieve easier operability and practical application".[13] From the first convention for global language in Schemmerberg in 1882, with seventy participants, to the 1884 Volapük Congress in Friedrichshafen, to the III World Congress five years later in Paris, this artificial language experienced a dynamic and unprecedented development. Yet its ascent was simultaneously its downfall. Even as Schleyer composed the three-stanza

Volapük anthem (instructions: "Quietly solemn, for male voice"),[14] voices were raised within this choir of supporters who criticized the difficulty in learning the system and its unsuitability for aesthetic forms of expression. There were also difficulties in speaking Volapük, while official estimates for learning it were unrealistic at only a few days and, indeed, only a few hours under the tutelage of a world language instructor.

Schleyer's response to such calls for simplification or revision was naive. With the authority of the inventor, he rejected any and all modifications of the linguistic conception. As fast as it attracted attention, then, the first decidedly global language disappeared, leaving behind all manner of manifestations, such as the typographically nonsensical poems of Christian Morgenstern. Of the 283 Volapist societies once in existence, only 4 remained after 1900. However, the bold supporters of the idea of a global language had long since set their hearts on such a project and in some cases simply rededicated the title and prime occupation of existing societies to their new favorite: Esperanto, drafted by the Polish ophthalmologist Lazar Ludwig Zamenhof (1859–1917).

In 1887, Zamenhof published his variant of a global auxiliary language, Lingvo Internacia, under the pseudonym of Dr. Esperanto (the hopeful). Zamenhof, like Schleyer, lived at a linguistic crossroads: in Białystok and Warsaw, where Polish, Russian, German, and Yiddish collide. To remedy everyday linguistic confusion he devised an *a posteriori* artificial language modeled predominantly on several major European languages. For ease of learning, Esperanto has a comparatively small set of rules, which can be summarized on a single page, and it leans even more than Volapük on the vocabulary of natural languages, having around 75% Romance and 20% Germanic morphemes—thus hinting at the target linguistic communities served first by a global artificial language.

While Volapük, despite attempts at institutionalization such as the establishment of an academy, went from one crisis to the next and was limited in its spread mainly to grammar school teachers and mid-level employees in Central Europe, proponents of global languages increasingly gathered under the banner of the Esperanto movement, which had taken as its symbol a green five-pointed star, representing the continent in the color of hope. Shortly after the turn of the century, numerous Esperanto associations were founded in various countries: first in the land of the Universal Postal Union, Switzerland (1902); then in Spain and Mexico (1903); in England (1904); in the US and Bolivia (1905), and finally in Germany (1906). The international Esperanto Congress in Boulogne-sur-Mer in 1905 served as a prelude to the sustained series of yearly international gatherings, which continued until the outbreak of the First World War. As was logical, the national associations eventually consolidated into a World Federation in 1908.

Following the aims of the architects of planned languages, Esperanto not only serves as a *lingua franca* for daily use. In contrast to Volapük, it also has aesthetic and literary potential. Zamenhof himself encouraged the development of such potential by translating Shakespeare's *Hamlet*, Gogol's *Revisor*, Goethe's *Iphigenie auf Tauris*, Schiller's *Räuber*, and Andersen's fairy tales, as well as the Old Testament, poems, and proverbs. Aware of Couturat's suggestion of approaching Goethe's notion of world literature, other supporters of Esperanto added the *Ilias* and the *Monadologie* to the canon as a means of demonstrating both the language's capabilities, the "wonderful ductility" of Esperanto, and also to achieve the highest impact to promote its coming use.[15] The translation of world literature into a single global auxiliary language, it was generally agreed, would contribute to the wider dissemination of these texts as well as avoiding their translation into other languages, seen as a "frivolous squandering" of time and energy.[16] Couturat connects this claim to a media-theoretical argument which underscores the secondary character of an artificial language and which recovers the potential of the inherent dynamism of the original: "Global language will relate to the national literatures as photographs to the original works after which they were produced".[17]

Nevertheless, the development of Esperanto in some ways mirrored that of Volapük. The first Esperanto World Congress in Boulogne-sur-Mer in 1905 resulted in a split into an orthodox fraction, with a quasi-religious reverence for the original version of the language (the so-called *fundamento*), and a reform movement. Reformists emerged out of conspiratorial activities which demanded a drastic departure from the *fundamento* and, in 1907, following a gathering of the Délégation under the leadership of Louis Couturat, decided to form an autonomous language called Ido (in Esperanto, "descendant", "successor") as a derivation of Zamenhof's planned language. Initially much less autocratic than Schleyer, Zamenhof allowed a vote regarding possible changes to Esperanto in 1894. On the opposing side, Louis Couturat and Louis de Beaufront (elevated to the status of official creator of Ido) were joined by Wilhelm Ostwald, an eager advocate of basic improvements. Ostwald became the mouthpiece of the reform movement by not only leading the following meetings of the Ido committee, but also producing a flood of pamphlets in order to make the language public and foster conversion amongst old followers of Esperanto. Ultimately, Esperanto emerged as strengthened by the linguistic conflict, particularly as the frequent modifications and revision of Ido contributed little to its linguistic stability. By contrast, Esperanto blossomed in the interwar period, even if its dissemination remained severely limited. Indeed, historians of Zamenhof's world auxiliary language characterize his construct today as the "means of communication of a world-wide diaspora community".[18]

As for the aftermath of the Ido project, Ostwald and Couturat cooperated closely—at least initially—in their endeavors to institutionalize Ido by establishing their so-called *World Language Office* in 1910 and the Initiative Committee for the Establishment of an Association for the Creation of a World Language Office, which in fact never began operations. Consequently, the Ido Academy saw no further appreciable success. As a result,

Ostwald's efforts for a world auxiliary language, rather than progressing, maintained a state of permanent tentativeness. After 1915, his preoccupation with the advantages of internationalism come to an abrupt end as he embarked on a new project: *Wede*.

If Ostwald's 1903 lecture to the Bavarian regional association of the German Association of Engineers (VDI) served as the prelude to his intensive engagement with international auxiliary language projects, then his speech at the end of October 1915 marked the end of this preoccupation. Ostwald now sought to promote a new planned language under the name *Wede*, one which had nothing in common with the harmonizing and pacifist air of previous concepts. If one previously had to ask to which nation Ostwald's industrious world official was obligated, this Monistic Sunday Sermon provides an unequivocal answer. The address reveals nothing less than the conversion of the notion of the world language from Paul to Saul:

> I propose to produce a simplified German on a scientific-technical basis for practical use in those areas [i.e., the newly occupied countries]. In this, all dispensable variations, all of the aesthetically charming richness of the language which complicates its learning so tremendously, must be removed, so that this new means of communication, for which I propose the name Weltdeutsch, can be learned and used by everyone with little effort.[19]

Thus *Weltdeutsch*, or *Wede* for short, became Ostwald's principal focus.

What began in 1901 with a Délégation and an ambitious program to establish a single auxiliary language throughout the entire world deteriorated under combative conditions into a neglected political farce. It is worth noting that this collapse took place precisely under the title *Weltdeutsch*. In Ostwald's internationalist understanding before the war, "*Welt*" professed to denote absolutely everything, but was nevertheless completed with the addition "-deutsch", under the influence of the Schlieffen Plan, as well as a supposedly successful eastward expansion. It was the World War which first revealed the hegemonic impetus of Ostwald's series of world projects, carefully hidden behind euphoric terminology. "*Welt*" had denoted the largest possible target area of an undertaking which, understood without further consideration to be homogenous, was eagerly anticipating change and was to some extent malleable.

Even early critics of the movement called attention to the extent to which this notion was pervaded by the methods of imperial politics and its colonial influences. The "world" of a world language had never been "understood in that sense as a label for the entire inhabited planet".[20] Meyer sharpened his argumentation and turned it into a broad, substantial objection: "One may not abandon oneself to illusions regarding the fact that the overwhelming portion of all the inhabitants of our planet has not the least interest in the creation of a world language".[21] The apologists, or perhaps "apostles", of this world auxiliary language nevertheless seemed to take such criticism as a challenge to win over the majority. With such arguments, the global auxiliary language movement became from the beginning a kind of mission predestinated to bring a message of salvation to their "new pupils", the linguistic "heathens" of the whole world. However, other means were available for the delivery of the "gospel" in a media-technological planet understood as a battlefield ready to be conquered by a linguistic project. The roles of sender and receiver remain unquestionable:

> Steam shortened distances; electricity neutralized them. From this ensued an intensive and extensive development of industrial and commercial relationships between all peoples. The civilized world, which was limited to old Europe a century ago, has incorporated new nations and entire portions of the world. The European market stretches across the entire globe.[22]

What, then, does "world" mean from the perspective of the artificial language movement? Couturat characterized "world" as the progressive incorporation of other continents through the institution of Europe and its market power. The advocates of world language planned to launch their rational program of a unifying language through distribution and communications channels, beyond any obvious political claims to power. The global auxiliary language project grafted itself onto the global transport and communications system controlled from within Europe. However, at precisely the moment in which Europe threatened to be torn apart by war, the apostles of world language was forced to recognize that their medium of transmission, world communications, was suddenly bogged down and, in places, was coming to a complete standstill. Additionally, they had to acknowledge that their supposedly neutral linguistic program pursued a massive hegemonic aspiration: their project had been (geo-)political all along. The battle for world language, to quote Gustav Meyer again, "will be connected at the deepest level to the political configuration of the Earth's surface".[23]

Thus, the specter of a completely accessible world necessarily collapsed at the moment when the bastion of Europe proved to be split, when the World War dispelled the illusion that a world language could be established through the argument of plausibility or wasted energy, by exposing the actual power relations and mechanisms of power of linguistic policy. The innocent dream of international institutions such as the World Language Office receded into the distance. With war reigning in Europe, the "circle of the civilized world" constricted still further. The basis for the world language movement narrowed appreciably, splitting along with its spokesmen into national factions; a new schism threatened, this time between Germany and the *entente cordiale*. Couturat died in 1914 in a traffic accident, and the notion of world language, for want of an alternative or an antagonist, struck up a new alliance with German. Through Ostwald's sermon, in which he transformed himself into an anti-internationalist, he also officially adopted Meyer's insight that the spread of a world language was "most certainly a question of power".[24] Ostwald recommended the practical use of *Weltdeutsch* "first in those areas" which were already subdued—and "subsequently in the whole world". Meanwhile, Ostwald still clung to his idea of projection, of sending out the message which promised to lead to success, that is to say, to maximum extension. However, instead of global communications, the main media-technological conditioning factor was now that of Germany. The unifying notion of "world" in the sense of internationalism was shattered; yet the *Welt* of *Weltdeutsch* still denoted the old, all-encompassing target area, and the notion of *one* standardized world had now lost its political innocence.

NOTES

1. Markus Krajewski, *World Projects. Global Information before World War I* (Minneapolis: University of Minnesota Press, 2014). All translations from the German sources were made by Charles Marcrum II and published in this work, unless stated otherwise.
2. Wilhelm Ostwald, *Energetische Grundlagen der Kulturwissenschaft* (Leipzig: Verlag von Dr. Werner Klinkhardt, 1909), 171: "jedes Mal ein tiefes und lebendiges Weltheimatsgefühl, wenn man bei solchen Gelegenheiten mit Männern zum ersten Male zusammenkommt, deren Leistungen man aus der Literatur kennt, und wenn man sich mit ihnen nach den ersten Begrüssungsworten alsbald auf gemeinsamen, beiderseits gleich vertrautem Boden begegnet".
3. Wilhelm Ostwald, *Die Forderung des Tages* (Leipzig: Akademische Verlagsgesellschaft, 1911), 453: "Da standen die Männer nebeneinander, die sich gegenseitig das Belangreichste zu sagen hatten, aber sie konnten sich nicht verständigen. Denn wenn die meisten Gelehrten und Praktiker heute auch mehrere Sprachen soweit beherrschen, dass sie Fachabhandlungen *lesen* können, so ist es doch von diesem Punkte noch eine weite und mühsame Reise zum *mündlichen* Verkehr in der fremden Sprache. So entstand aus der Not der Gedanken der internationalen Sprache von neuem".
4. Johann Heinrich Gottlob von Justi, "Gedanken von Projecten und Projectmachern" in *Politische und Finanzschriften über wichtige Gegenstände der Staatskunst, der Kriegswissenschaften und des Cameral- und Finanzwesens*, ed. Johann Heinrich Gottlob von Justi (Koppenhagen, Leipzig: Auf Kosten der Rothenschen Buchhandlung, 1761), 257: "Wohlfahrt ganzer Reiche, ja der ganzen Welt".
5. Louis Couturat, *Die internationale Hilfssprache* (Berlin: Möller & Borel, 1904), 4: "Was nützt es, dass wir von einem Ende der Welt zum anderen reisen, schreiben, sprechen können, wenn wir uns nicht verstehen?"
6. Couturat, *Die internationale Hilfssprache*, 5. "die zweite Sprache für Jedermann".
7. Couturat, *Die internationale Hilfssprache,* 6. "Die Weltsprache soll . . . keine technische oder aristokratische Sprache sein, die nur einigen Eingeweihten zugänglich ist, sondern eine tägliche Sprache, die ihre Anwendung auf der Eisenbahn und im Gasthofe ebenso findet, wie in den gelehrten Gesellschaften und auf den Congressen".
8. Wilhelm Ostwald, *Die Organisation der Welt* (Basel: Verlag des Weltsprachevereins Ido, 1910), 12: "Die Wahl irgendeiner natürlichen Sprache ist ausgeschlossen, aus dem einfachen Grunde, weil sie alle nichts taugen".

9. Couturat, *Die internationale Hilfssprache*, 6f.
10. Friedrich Nietzsche, *Menschliches, Allzumenschliches I* (Munich, Berlin: dtv/de Gruyter, 1999), 222: "und in irgend einer fernen Zukunft wird es eine neue Sprache, zuerst als Handelssprache, dann als Sprache des geistigen Verkehres überhaupt, für Alle geben, so gewiss, als es einmal Luft-Schiffahrt giebt. Wozu hätte auch die Sprachwissenschaft ein Jahrhundert die Gesetze der Sprachen studirt, und das Nothwendige, Werthvolle, Gelungene an jeder einzelnen Sprache abgeschätzt?"
11. Ostwald, *Die Forderung des Tages,* 452: "Versammlung zur Feststellung internationaler elektrischer Einheiten".
12. Ostwald, *Die Forderung des Tages*, 452: "zuletzt mit staatlicher Autorität für den allgemeinen Gebrauch zugänglich".
13. Rupert Kniele, "Der erste Kongress der Weltsprachefreunde," in *Der Erste Volapük-Kongreß, Friedrichshafen, August 1884. Dokumente und Kommentare*, ed. Reinhard Hauptenthal (Saarbrücken: Artur E. Iltis, 1984), 9: "der *Weltpost*, dieser grossartigen Errungenschaft der Neuzeit, zur leichteren Durchführbarkeit und eigentlichen Geltung zu verhelfen".
14. Reinhard Haupenthal, *Der erste Volapük-Kongreß, Friedrichshafen, August 1884. Dokumente und Kommentare* (Saarbrücken: Artur E. Iltis, 1984), 63. A sample from the "Hymn of the Volapists", second stanza:

Kìs alsò kanòs koblòden	Tell me, what can increase
lelikà volà menìs?	the highest good of every man?
Kìs alsò kanòs menòden.	A single tongue will bring salvation
as pak bàl. omsà stadìs?	if it sounds in every land.
Klù tonòdosèz in val:	Therefore let the whole world know:
mènade balè pak bàl!	a single tongue from pole to pole!

15. Louis Couturat, *Eine Weltsprache oder drei? Antwort an Herrn Professor Diels* (Stuttgart, Leipzig: Deusche Verlagsanstalt, 1907), 15: "wunderbare Bildsamkeit".
16. Couturat, *Eine Weltsprache oder drei?* 14: "frivole Verschleuderung".
17. Couturat, *Eine Weltsprache oder drei?* 15: "Die Weltsprache wird sich zu den nationalen Literaturen verhalten wie die Photographien zu den Originalwerken, nach denen sie hergestellt sind".
18. Alicja Sakaguchi, "Die Dichotomie 'künstlich' vs. 'natürlich' und das historische Phänomen einer funktionierenden Plansprache", *Lingvaj problemoj kaj lingvo-planado* 20 (1996): 18. "Kommunikationsmittel einer weltweiten Diaspora-Gemeinschaft".
19. Ostwald Wilhelm, "Weltdeutsch", *Monistische Sonntagspredigten,* no. 36 (1915): 557: "Ich schlage vor, für den praktischen Gebrauch zunächst in jenen Gebieten ein *vereinfachtes Deutsch* auf wissenschaftlich-technischer Grundlage herzustellen. In diesem müßten alle entbehrlichen Mannigfaltigkeiten, all jenerfür die Ästhetik so reizvolle 'Reichtum'Reichtum¿keiten, all jenerfür dr Grundlage herzustellen. In diesem müßten alle entbehrlichen Mannigfaltigkeiten, all jenerfür d' der Sprache, welche ihr Erlernen so ungeheuer erschwert, beseitigt werden, so daß dieses neue Verkehrsmittel, für welches ich den Namen *Weltdeutsch* vorschlage, vorn jedermann mit leichter Mühe erlernt und gebraucht werden kann".
20. Gustav Meyer "Weltsprache und Weltsprachen", *Essays und Studien zur Sprachgeschichte und Volkskunde* 2 (1893): 38: "in dem Sinne einer Bezeichnung für die ganze bewohnte Erde zu verstehen".
21. Meyer, *Weltsprache und Weltsprachen,* 43: "Man darf sich keinen Illusionen darüber hingeben, daß der überwiegende größte Theil aller Bewohner unseres Erdballs an der Schöpfung einer Weltsprache nicht das mindeste Interesse hat".

22. Couturat, *Die internationale Hilfssprache*, 3: “Der Dampf hat die Entfernungen abgekürzt; die Elektricität hat sie aufgehoben. Daraus ist eine ebenso intensive wie extensive Entwicklung der industriellen und Handelsbeziehungen zwischen allen Völkern erfolgt. Die civilisierte Welt, welche sich vor einem Jahrhundert auf das alte Europa beschränkte, hat sich neue Nationen und ganze Welttheile einverleibt. Der europäische Markt erstreckt sich über die ganze Erde”.
23. Meyer, *Weltsprache und Weltsprachen*, 40: “wird mit der politischen Gestaltung der Erdoberfläche aufs engste zusammenhängen”.
24. Meyer, *Weltsprache und Weltsprachen,* 40: “auf jeden Fall eine Machtfrage”.

13 Impacts of a Global Language on Science

Are There Disadvantages?

Scott L. Montgomery

> *Men have no right to put the well-being of the present generation wholly out of the question. Perhaps the only moral trust with any certainty in our hands is the care of our own time.*
>
> Edmund Burke

During the summer of 2010, I spent three weeks in northwestern Australia as part of a geology seminar run by the University of Washington. Our travels took us through a sizeable portion of the Kimberleys, a wild and remote region of low-ridge mountains, sandstone gorges, and small towns encased by vast savannah ranches, bordered by aboriginal communities. These communities often combine people from different tribes, whose original languages have grown weak or gone extinct. One evening, in the town of Timber Creek, I fell into conversation with Roger, a man perhaps in his late thirties. It emerged that he was a father of two boys, just as I was, and so we first traded tales of sibling relations and discipline problems. Then I mentioned the language issue. His boys, he said, knew three languages, none of them completely. Roger's birth tongue was Warimajarri, and he also knew Gurindji pretty well, but his wife came from over Tennant Creek way and spoke Warumungu and also Warlpiri. Both parents knew Kriol, a pidgin that blended English and several aboriginal tongues, used by white settlers and natives since the early twentieth century as a *lingua franca* across much of the region. The boys spoke their father's language fairly, their mother's first tongue, Warumungu, better still, and Kriol best of all, because both parents and the extended family used it. Roger noted it was common for his sons and their friends to switch back and forth between all three languages, depending on who was talking and whether or not they wanted any adult to know what they were saying. A big problem, he said, was that there are a lot of English words in Kriol. "The whitefellas always thought me boys could speak English fine, but they can't. They had terrible time in school. Now the government change everything in school to English, so I think they learn it better". I asked him how he had learned his English. "I learned well at school—me mum made me study hard! (laughter). Later, I worked at a cattle station. Had to use it there. Now I been working in a store about ten year, using English every day". It was time for me to get back to camp, so I shook his hand, wished his family well, and began to leave. "Hang on", he said, suddenly. "Your boys in school, right? How many tongues they speak?" I said, "Just English for now, though they're studying another in school". He flashed a smile and shook his head: "Got to do better, mate! One never enough!"

1. INTRODUCTION

At the level of international communication, English has gained a profound dominance in the natural sciences. There is every indication that, as a global *lingua franca* for science, English will not only remain in place but continue to expand further, particularly as more of the world's nations develop economically and expand their own scientific and technological capability. Every field of technical endeavor, where it aims at a multinational audience, has seen its journals, books, conferences, symposia, lectures, workshops, and internet usage shift to the English language over the past several decades (Montgomery, 2013).[1] Measures that employ the largest bibliographic databases for the sciences (e.g., Web of Science, Scopus, Chemical Abstracts, PubMed) lead one to conclude that as much as 90% or more of international scientific publication now takes place in this language for a wide variety of disciplines, from physics to medicine.[2] Criticism that some of these databases are biased toward English appears legitimate,[3] yet it does not seem to matter, since studies of the literature in individual fields and subfields and of such things as conference papers from international meetings give the same overall result. So far has English penetrated the sciences that its use is now expanding for domestic publications and conferences in non-Anglophone countries.[4]

It is common to extol the benefits of a global tongue in science. Such benefits are perceived to include not merely those of a short-term practical nature, like expanded collegiality, easier collaboration, and broader and more immediate dissemination of findings, but advantages that will greatly profit science in the long-term through the globalization of knowledge that, once fully realized through the medium of the internet, will help create a "collective intelligence" with the power to "reinvent discovery" itself.[5] There is, too, the symbolic capital of a "global scientific community" embodied in a global scientific language. Most researchers around the world, therefore, if asked to comment, would be unlikely to find the status of English problematic or controversial.

Yet English is by no means the first world-scale language of science, and any close look at such languages of the past will show they have not always been kind to their foreign users. First of all, these former tongues of science—Greek, Latin, Arabic, Persian, Chinese among them—did not attain authority in consensual fashion, but arrived on the back of conquest and empire-building. The impacts they had, for example, on tongues that existed before their arrival, were often mixed. They could bring extinction for native languages in conquered territories, but also, over time, the revitalization of others and the creation of new tongues as well, such as the dozens of Romance languages that eventually emerged from Latin or the multiple forms of Arabic across North Africa and the Middle East. Second, a dominant language has historically altered many institutions of scholarly practice—schooling, literacy, practices of reading, the technology of writing, the structure of intellectual communities, even the role of orality versus textuality, have all been forced to change. A complex process of adapting institutions and practices to the new language, while also adapting the language to local needs and uses, has been the rule.[6] Overall, adaptation has involved struggle and loss, as well as gain. There are thus limits and drawbacks to be considered when a powerful *lingua franca* gains authority.

To what degree might this be the case with English in the natural sciences? Are there important disadvantages that stand out, and, if so, how serious are they both for the present and the future? Given the many years of training and the intense competition for resources and rewards in contemporary science, can any such problems be addressed in some way? In truth, these are central questions for the future of scientific endeavor in terms of researchers themselves and their social practices. They have been addressed by an expanding literature dominated by scholars of language who tend to view the global role of English in strongly negative terms.[7] Questions about fairness, access to knowledge, linguistic diversity, and more are regularly posed. It is essential, then, to review and interrogate the principal drawbacks that have been identified.

2. THE ISSUE OF EQUITY

A global tongue seems to cleave the world in two—into those who *can* and those who *can't*. In truth, of course, the situation is more complex: there are those who *can* by native advantage; those who can from years of disciplined or privileged study; those who partly can, due to limited training, and are striving to improve; those of an earlier generation who learned a different world language in their youth, e.g., Russian or German, and now must learn English; those who feel they must forever struggle in a gray zone of inferior ability; and those who have little or no training in the tongue of power and are thus left outside the wall, whether by choice or lack of chance. Each of these groups is very much present with regard to English in the natural sciences today.

Though a world language provides new opportunity for scientists in a host of ways, it brings challenges and barriers as well.[8] This was stated some time ago by Eugene Garfield, founder of the Institute of Scientific

Information and its Science Citation Index (now one of the world's largest bibliographic databases):

> The fact that English is the internationally accepted language of research communication raises the issue of a language barrier in two senses. First, those whose native language is not English risk being unaware of—and overlooked by—mainstream international research unless they learn to read, write, and publish in English. Second, native English-speaking researchers risk being ignorant of significant findings reported in foreign languages . . . unless they become proficient in at least one other language.[9]

The world according to Garfield is one where non-Anglophone researchers *must* learn English and Anglophones *probably should* learn another tongue.[10] Nowhere does a concern about fairness or "linguistic injustice" appear.

It is an issue that cannot be ignored: mother tongue speakers of English derive advantage from this fact, without doubt. Speakers with English as a foreign language, or a fair portion of them, must spend time and effort learning it and perhaps paying to have their papers edited and corrected. Meanwhile, the specific benefits for mother tongue speakers are far from trivial. Anglophone researchers are entirely at home in the required discourse—they face the fewest obstacles and risks in using and producing it and so can more easily find entrance to its benefits (though this is never guaranteed), which include the rewards system in science. Domestic journals in the U.S., U.K., Canada, or Australia become international by default. Native-like competence can, in some cases, aid the likelihood of getting a paper published, or having it cited by others, thus improving one's standing and professional prospects.

It is important to realize, however, that the same advantages hold for any and all fluent or highly competent speakers of English. Nations where English is successfully taught as a basic skill, where a large portion of college graduates are proficient, fall into this category. They include, for example, not only most former British colonies, where English can be a second or at least official language, but also all the Scandinavian countries and the Netherlands. Strong levels of proficiency are also achieved in Austria, Belgium, Germany, Switzerland, and Malaysia.[11] While there are no official figures for the number of researchers worldwide who are highly proficient in English, it seems likely that the total is significant. Certainly it has been rising rapidly overall since the 1990s, according to recent statistics.[12]

But for speakers of English as a foreign language (EFL), with less skill and confidence, the noted benefits can be wholly inverted. Writing and presenting can require a great deal of extra work, and also money if texts are sent for outside correction. Some journals will undertake such editing themselves, if they perceive the article worthy enough, but a majority lack the resources to do this and so may well recommend it receive outside help. Effort by the author(s) is therefore drained away from other tasks like research itself (though communicating is essential scientific work too). When added to the economic challenges commonly faced by scientists in poorer countries, these demands can seem discriminatory.

There are subtle hurdles as well. Linguistic conventions in scientific writing can differ both slightly and significantly among languages. Verb tenses, use of adjectives and qualifiers, the organization of a paper, and repetition of points are all examples. If science in French commonly employs the present tense (rather than the past tense, as in English), Japanese technical discourse is highly prolix (by English standards), while Chinese papers may include "re-use" of wording from published material.[13] All of these things increase the risk of error, misunderstanding, and transgression for texts by non-native speakers. On the other side, some Anglophone readers may view an article with non-standard ("odd" or "improper") vocabulary or awkward ("incorrect") phrasing off-putting or marred, despite worthy data. In conference talks, poor or imperfect English, heavily accented or mispronounced, can frustrate listeners who expect only near native fluency and may lead some of them to associate imperfect performance with low importance.

There are social effects to consider. Researchers who are more competent in English, who have studied and worked in Anglophone nations, can find better job opportunities to the point of gaining positions of authority in a nation's research establishment. Competence in English, perhaps even more than competence in research itself, may thus become a criterion for positions of power in a wide spectrum of scientific institutions, enhancing the status of the language still further. English skill is routinely interpreted as the ability to connect with, and utilize, international science and all it can offer. Such is logical in one sense but will appear discriminatory as well. Beyond this lies the global need for language teaching and translation; nations with mother tongue and fluent speakers of English gain multibillion-dollar industries in these two areas and so become, through their teachers and translators, those who have some direct influence over the flow of language and the movement of knowledge within it.

Issues of fairness therefore emerge not from the brains of academics but from the real world. They can be disputed in terms of their true extent and ultimate effects, but not in terms of their relevance and actuality. They do, however, raise questions of their own, and these should not be avoided or dismissed either. Here are the major ones:

1. How common, in actual terms, is bias against non-native speakers of English in international scientific publication? Is it routine that reviewers weight imperfect English more heavily than other criteria? Some statements by editorial board members suggest this may not be the case.[14]
2. If the advantages attached to English now extend to any individual who is highly proficient, native speaker or not, what does this imply about reducing the fairness problem elsewhere?
3. The history of science shows that in eras where a *lingua franca* held sway, intellectual culture did adapt, though time was needed for this

to happen. Could the fairness problem represent a temporary phenomenon, something that will be largely reduced over the next few generations?

4. Because any global language for science would bring the same issue of fairness, what is the preferred option? Would it be more desirable for scientists to have several international tongues to choose from—say, Chinese, Spanish, English? Would not a majority of future researchers around the world still need to learn a foreign tongue?
5. How true is it to say that scientific advance in English is wrongly equated with scientific advance in general? Is this entirely erroneous or just a crude approximation? How often are national findings re-published in the international literature? If researchers in all of the world's major scientific nations now employ this language to publish in international journals the findings they believe are most important, can it not be said that the major part of scientific progress is now shared with the world in English?

The fairness issue, therefore, is not at all simple, and it raises difficulties of its own. It cannot be factored down to a doctrine of victims versus victimizers or a conspiracy among Anglophone editors or a matter of "strengthening U.S. global hegemony". Such characterizations are shallow, unhelpful, and usually wrong. Thoughtful critics understand and accept the status of English in science and the advantages it brings, yet they also look for ways to address the inequalities it creates. They have suggested an array of "solutions". By far the most common of these is to have journals provide financial or translation support for EFL authors.[15]

Journals, as well as academic departments and other institutions, that can afford to help EFL speakers with their manuscripts (and with advancing their English skills) would be doing good service, for themselves and for science. This is clear; it defines an investment in human capital, one with inevitable returns. Yet we should not be sanguine about what is involved. Improving "troubled" manuscripts is labor intensive, akin to translation. Most periodicals in international science are now owned by for-profit corporations, and so convincing companies like Elsevier, Kluwer, or Taylor & Francis to invest in science with a capital "S" might prove challenging.

3. CONFINED TO THE MARGINS

Related to fairness is a second issue: the power of a global language to marginalize other tongues and their speakers. This can be as much a personal feeling on the part of individuals as an institutional matter of policy. Lack of confidence or a sense of burden and insufficiency can be a dam to productivity for researchers engaged in high pressure, competitive work.[16] A scientist who feels insecure about her or his English ability can be less willing to share work with the international community. She may be an excellent researcher yet reluctant to write articles, give lectures, or speak at conferences and other gatherings where new findings are first revealed and crucially vetted by verbal exchange. If forced to participate by the rules of his institution, he may spend many long hours on a paper or talk, taking valuable time away from the lab or field (or both). He may seek collaborators whose English is better, even if they were not integral to the relevant research, thus diluting his or her own apparent contribution in any publications. She may even engage a ghost writer or editor who will help a good deal but leave her no more capable and no less unself-confident than before. Again, since research faculty are often evaluated by quantity of publications and their "impact" (level of citation by other researchers), any disincentive for submitting papers to international journals and symposia will have its own detrimental effects. All of these factors on a personal level—as both survey and anecdotal evidence suggests—can give scientists the sense that they are operating more on the margins of their field than the mainstream or the productive frontier.[17]

These types of experiences take place within a larger context. The global scientific community can now be roughly divided into three domains: 1) countries wealthy enough to support all fields of science or a large selection of them; 2) nations whose overall research capability is growing fast, but who must still make choices about what fields to fully encourage; and 3) developing nations able to support only a few fields of research, which tend to be applied science related to development goals of the country (agriculture, water, energy, public health are mainstays). The first group, no longer confined to Western nations plus Japan, has come to include countries like China, Brazil, India, and South Korea. The second group, by far the largest and most diverse, is led by states like Mexico, Turkey, Taiwan, Estonia, and Poland, most of whom over the next few decades will likely advance into the first group or arrive at its border. Finally, the third group comprises many of the least developing countries, where even basic services like electricity and clean water are not yet widely available. Nations in the second and third groups work with limited resources and scientific infrastructure; they remain therefore unable to participate very deeply in fields where Big Science is the rule—high energy physics, nanotechnology, genetic research, and large areas of engineering and medicine. Scientists in these countries can find themselves on the periphery of contemporary research publishing almost by decree, faced with inadequate funding, inferior technical training, and overburdened facilities. Demands for use of English can therefore add yet another layer to this structural context of marginalization. Such is confirmed by discussions with authors in developing countries, who list this as among the problems they face for publishing in international journals. The concept of a global program to offer editing help to authors in poorer nations (the third group, as listed earlier) therefore seems something of considerable worth.[18]

Yet it would not solve the structural problem. Serious limits to scientific capacity, to computer and broadband technology, for example, will place any research program well away from the core of cutting edge work—work that would be welcomed by international journals. Even were it to be put into English, the research of scientists who suffer from such limits would likely be considered less worthy in direct competition with that performed at better equipped facilities in more advanced countries. English is involved here as well. Language programs in group 3 countries, being fairly new or still rudimentary, often do not have enough qualified teachers and are less effective. Young researchers, not well taught in English, have a definite handicap in applying for both national and international grants, even when the money is specifically for their ranks. Taking Vietnam as an example, these problems are not helped by the truth that young scientists who long for high level training abroad (and who might serve as scientific leaders upon their return) are often ineligible because of poor English skills.[19]

Demand to use English can intimidate many authors in group 1 nations, too. It is not hard to imagine the following scenario repeated in a number of countries, particularly in Asia:

> The nervous [Japanese, Korean, Chinese, Vietnamese, Thai] postdoc spent two weeks creating slides, 30 hours drafting a script and 44 hours rehearsing. Altogether, she spent one month away from the bench so that she would not disappoint her supervisors and colleagues during a short informal presentation, in English, before her co-workers. Yet they remembered only the mistakes, she says.[20]

Our compassion is surely brought to bear. We readily conceive an entire universe of small, daily tragedies like this. Yet a moment's reflection may be needed. That the presentation is "informal" and for co-workers tells us English is an integral part of the scientific culture here (every postdoc must do presentations as training for upcoming conferences). Some members of the research team are proficient, if demanding, and expect improvement. Doing science at this institution means using English, a baseline condition, understood by all. In this light, the situation becomes less draconian, more historical. Might we then, in this scene, be looking at a single (larval) stage in a longer evolution? Anecdotes about struggles and "failures" are not the only stories to be told. How did the proficient ones get that way? Moreover, we should not overlook the small fact that this "tragedy" takes place in one of the richest, most scientifically vibrant nations on Earth (Japan). Saying that this proves how difficult the English situation must be misses the point. In a wealthy nation, with a high level of education and scientific instruction, becoming competent in a second language cannot be counted an overwhelming barrier. The same situation, after all, would arise no matter what foreign language needed to be acquired.

A greater tragedy happens for those scientists who are capable in English but marginalized due to economic and political factors. Compared with the Japanese or Brazilian biologist who has the advantage of excellent training and a good salary, top facilities, and private sector R&D of the highest order, a researcher in Moldova or Bolivia has little of any of this, whatever their level of language competence. They therefore find reasons for leaving their home country, and not merely because of the role of English as a global tongue. Indeed, a global language may well provide one of the only pathways for researchers who stay in their countries to collaborate, gain notice, and participate in the larger scientific community.

Realities and assumptions surrounding the issue of marginalization therefore also beg some level of interrogation. The major questions we might ask would seem to be these:

1. Are complaints about lack of confidence regarding English really more about the quality or extent of language learning? Some countries who teach English to their children from an early age have highly successful programs (Norway, Netherlands), while others do not (Spain, Japan). Should more attention be given to the reasons for such disparity?
2. Is it not legitimate to expect MS and PhD students, who have acquired highly complex knowledge in challenging fields of science, to be able to achieve competence in English?
3. Would it make sense to include English in scientific training as a required subject, like mathematics? Or would this prove yet another unfairness, taking time away from other technical subjects?
4. Developing nations have shown rapid growth in English publications between the 1990s and 2000s. Obviously, *some* scientists in these countries have become competent enough to use English regularly. As with the issue of fairness, might it be that the problems of marginalization represent a temporary stage in the historical fulfillment of a global language in science?
5. Which, in the long-run, is likely to be more marginalizing to a nation's research—the struggle to adapt to a global language or the refusal to do so?

4. BIAS OR STATUS?

A global language grows, in part, by status momentum. The very perceived importance of such a language magnetizes new speakers and also draws to itself institutional policies in its favor.[21] Attached to one or more powerful nations, with whom many forms of opportunity related to training, jobs, and publication are associated, the effect will be still stronger. At a certain point, official bodies, educational systems, and other mediatory sites take up the language, spreading it further, integrating it more deeply, compelling its use by future generations. Self-reinforcing cycles therefore come to exist. Institutions that deal in communication can play a particularly significant and unexpected role.

For science, one such institution has done just this. The bibliographic database Science Citation Index (SCI), created by Eugene Garfield in 1960 and now available in greatly expanded form as the *Web of Science*, covers over 6,650 of what are described as "the world's leading scientific and technical journals across 150 disciplines".[22] Besides publication trends, SCI famously ventured into areas of "scientific worth", mainly through the hotly debated "impact factor" (IF), a measure of the average citations per paper. By the late 2000s, citation data from SCI and now also a new competitor, Elsevier's Scopus, were being incorporated into an array of applications. In addition to North America and Europe, Japan, Korea, Taiwan, and especially China were now making use of citation data to evaluate their own science. Editors, scientist-authors, librarians, and hiring committees, as well as policy officials, were all paying attention to IF data. In a number of fields—medicine, above all—IF had become fully integrated into scientific publishing itself, bringing its own apparent impacts, with journals even manipulating their content to obtain higher ratings.[23] Researchers began to find themselves assessed and funded partly on the basis of this data (despite cautions provided by the Institute for Scientific Information), especially in Britain. Some five decades after it was born, SCI was no longer in the business of mere analysis; it had become part of the social practice of science itself.

Early on, ISI was convinced that English had become the dominant language of international science and would only grow in this role with time (see prior quote by Garfield). Such was not a mere matter of blunt and unapologetic bias. In 1971, the Institute decided to do an analysis of journal citation patterns. Using data from the last quarter of 1969, ISI created a list of the most frequently cited journals and found, to its surprise, that a mere 152 out of 12,000 journal titles accounted for 50% of all citations. A miniscule portion of international publication turned out to have massive importance. Of the 152, moreover, only 30 were journals from non-Anglophone countries (mainly the Soviet Union, Germany, France, Scandinavia, and Japan), with about half of these in languages other than English. When a second list of 152 was assembled for comparison, based on IF, the number from non-Anglophone nations shrank to 20.[24] It therefore seemed obvious to ISI personnel that an overwhelming majority of "high impact" journals were in English.

Critics maintain that SCI data, which has dominated discussions of "scientific output" until very recently, has been biased from the start in favor of English.[25] That SCI has claimed to identify the "top journals" in every major field, yet only selected those that are either entirely English or that contain important parts (abstracts, references, keywords) in this tongue, gives some credence to the claim. Even in the case of what are called "regional journals"—those, for example, in Spanish that have strong distribution across Latin America—SCI demanded that bibliographic elements be in English. When compared with other databases specific to particular fields, it is clear that SCI has historically been highly selective and likely under-represented the real activity of scientific advance. Its past refusal to regularly admit such outlets as conference proceedings and, above all, open access archives and online-only publications has been notable, though temporary. To an important degree, that is, SCI has been highly conservative, tying its view mainly to traditional hard-copy science and its online derivatives. No surprise that the breadth of its coverage tended to decline over time. Nor that it seemed weakest in some of the fastest growing fields that mainly employ the internet, such as computer science.[26]

The total result has been that all-English journals and researchers who write only in English have a far larger presence in the database. Without proper caveats, this can warp analyses of scientific publication, since in any nation with a scientific tradition there will be much incentive to publish in the native language as well. Authors who did so therefore could suffer from the fairness and marginalization issues, due in part to the influence of SCI itself. As its indicators have been put to greater use by those in positions of power over funding and hiring, it has had its own impact on the trend toward all-English publication. Small surprise, then, that Anglophone nations have appeared to be consistently over-represented in the results of various measures (e.g., papers per researcher, per university, per country).

Not all measures, however. When gauged in terms of articles per million inhabitants or per capita, rankings shift considerably. For example, over the period from 1998–2000, Switzerland was seen to lead the table using such measures, and the Scandinavian countries (Norway, Sweden, Finland, Denmark, Iceland) came out particularly dominant.[27] Meanwhile, IF scores calculated for the period from 2005–2008 also show Switzerland highest, followed by Denmark, with the U.S. third. Though the U.K., Ireland, and Canada were in the top ten, so were Norway, Sweden, Belgium, and the Netherlands.[28] Northern Europe plus Switzerland, therefore, emerge as the true global scientific leaders by this measure. Such offers a very different portrait of world science than the standard triad posing North America versus EU versus East Asia. Except for the U.K., these "new" leaders are the most multilingual nations in Europe—well-educated Swedes, Swiss, Danes, and the Dutch commonly know three, even four, languages, one of which is English.

Thomson Reuters has taken note of the criticisms against it and, under growing competitive pressure from other databases, has sought to broaden its offerings. Since 2005, it has expanded the number of non-English journals in its database so that its total coverage now extends to over 12,000 journals. It also now includes open-access journals, books, and conference proceedings—major increases in its total coverage. Moreover, it has sought to begin coverage of national science. In the latter case, it has formed a partnership with the Chinese Academy of Sciences (CAS) to create a separate citation database for Chinese publications. In 2011, this new database included 1,200 scholarly journals that the CAS helped recommend, with article information in both Chinese and English and about 40% of all papers carrying English-language abstracts. As a potential model for other countries, this implies not only the preservation of national science, selective though it may be, but also the creation of its visibility for the international community.

The bias problem therefore appears to be decreasing. It is a tempting thought that SCI, an institution presumably guilty of globalizing English in science, becomes a key preserver of the multilingual dimension to world scientific endeavor. Yet when we look at the larger picture of world science in the second decade of the twenty-first century, several questions present themselves:

1. Given the rapidly globalizing reality of research today, the fact that collaboration is more multinational than ever before, is the role of national science shifting in some way? Accepting that this domain of science remains crucial, how can we describe its importance as directly tied to national languages, beyond vague appeals to "linguistic diversity"?
2. Since other general databases in science, such as Scopus, and more field-specific ones like GeoRef, PubMed, BIOSIS, INSPEC, Compendex, and MathSciNet, show overwhelming dominance of English in the international literature, to what degree does the purported bias in SCI matter any longer? On the other hand, it might also be asked how neutral, or biased, toward English these databases might be.
3. Are there a significant number of researchers worldwide who complain of bias toward English and reject its global role in science today? It might be noted that researchers who have found fault with the use of IF and other "quality" indicators rarely, if ever, bemoan a pro-English slant in the database.

5. PLAGIARISM—AN OCCUPATIONAL HAZARD?

Another potentially serious issue is the lure to fraud, particularly plagiarism. Here is meant "plagiarism" as a problem of written language, specifically the copying of published text, not data or methods or ideas. For those who lack adequate English skills, the pressure to publish in this language can be said to encourage various types of "borrowing". It has been often observed, and lamented, that non-native speakers are over-represented in plagiarism cases.[29] Striking examples exist: a 2010 study of an English-language periodical in China, *Journal of Zhejiang University—Science* (A and B), indicated that as many as 31% of all papers published in a single period between 2008 and 2009 were plagiarized in some form.[30] While it is not clear how much of this may be related to the demand for writing English, informal discussions (my own) with numerous Chinese researchers suggest that it likely played a significant part. Major cases in other countries exist as well. One instructive example involved nearly 70 papers by 15 Turkish physicists (many of them graduate students) that appeared in an array of international journals published in the U.S., Poland, Germany, China, Czech Republic, Slovakia, and India. Blame has been attributed to a combination of isolation, the need to publish, and poor English skills.[31] At the same time, however, comments by one of the accused parties may offer an added dimension:

> For those of us whose mother tongue is not English, using beautiful sentences from other studies on the same subject in our introductions is not unusual . . . Borrowing sentences in the part of a paper that simply helps to better introduce the problem should not be seen as plagiarism. Even if our introductions are not entirely original, our results are—and these are the most important part of any scientific paper.[32]

However unacceptable, this may not entirely be dissembling. Studies of researchers and science students in various countries appear to indicate that the copying or fabricating of actual data is well understood as a major offence, while word-for-word use of another's work, particularly in sections of a paper that are introductory or merely descriptive, can be treated more as an "editorial matter". Such is not only true for East Asia, where the problem is well known and where students often treat such textual appropriation as "language re-use.[33] It can also be true for other regions, like Latin America.[34] Plagiarism, we might recall, is a distinctly modern, Western concept. From the medieval period up through much of the Enlightenment, copying written material from other authors, including well-known ones, was common practice.[35] In fact, the term "plagiarism" remains less than perfectly defined today.[36] Complete textual originality is what international scientific journals demand, but it seems not (yet) to be an uncompromising international standard everywhere. Universities in Anglophone nations may themselves be contributing to the problem, however inadvertently. By accepting undergraduate and graduate students with low-to-moderate scores on English qualifying exams, such as the TOEFL (Test of English as a Foreign Language), and then placing such students in an "English readiness" curriculum, these universities encourage a tendency in some of these students toward "survival" strategies for passing.[37]

Questions that deserve to be posed here therefore include the following:

1. To what degree is the global status of English the major factor here? Can the motive to plagiarize be more directly tied to structural factors related to institutional demands and authorial standards taught to students?

2. Might the problem of plagiarism reflect not only individual attempts at taking "short cuts" but larger, historical factors? For example, could it be related to the urgency of certain nations, like China and South Korea, to advance their scientific cultures as rapidly as possible, and the resulting demands placed on researchers?
3. Can such factors, in conjunction with an academic culture lenient toward plagiarism and still at an early stage of teaching English skills, more fully explain the problems seen?

6. ENDANGERMENT OF OTHER SCIENTIFIC LANGUAGES

English is no serial killer of the world's tongues, but it is the one language whose growing use in domestic research around the world can reduce scientific communication in other languages. It is the tongue that can replace national languages in this context. It is also the language where new standards are set in scientific discourse, especially where new vocabulary is coined (a central process in science), older terms questioned, and vestigial ones replaced. Moreover, English acts as the testing and proving ground for any new technical vocabulary created in other tongues—if a term is not adopted into English, its chances of survival are poor.

An international language in a particular sphere of communication thus works against linguistic diversity in that same sphere. This is known as "domain loss" in linguistic circles. It defines a phenomenon that has occurred innumerable times over the course of history. Indeed, all international tongues succeed by this process, to ultimately varying degrees. In Japan and Korea, for example, classical Chinese was chosen as the language of writing for centuries, until each country developed its own orthography, at which point it retreated mainly to the domain of written scholarship. Arabic, on the other hand, served a kindred purpose in Persia only to a limited degree (mainly in religious matters) and for a limited time. Languages installed by means of conquest have commonly taken over the domains of government administration, military matters, law, tax collecting, and some areas of engineering (e.g., road building). This has not at all meant the inevitable death of previous dominant tongues, as they have continued to thrive and expand in other domains, not least vernacular ones. Yet it has meant a certain loss for such languages, which are no longer able to satisfy every linguistic need of a community.

There is no doubt that the global role of English in the sciences puts pressure on other tongues to stay current. Researchers, editors, and other communicators in national science must make a choice whether to adopt new vocabulary directly from English, to search for an equivalent in their own language, or, finally, to use English itself. These choices, through the broader topic of domain loss, became the subject of heated debate in Scandinavian countries during the early 1990s, above all in Norway, Sweden, and

Denmark.[38] Behind the controversy lay a number of factors, such as background worries over national identity in the face of the European Union and, more specifically, ongoing linguistic questions in the public context, such as the struggle over standardizing Norwegian (a relatively new official language, since the country only achieved independence from Denmark in the nineteenth century) and achieving a single, understandable form of spoken Swedish. Another factor, however, was a fear of language endangerment—the concern that if linguistic territory were ceded to English in so important an area as research and training, all domains would eventually follow, and the national tongue would go extinct.

By the late 2000s, the controversy had largely cooled. Nearly twenty years after worries were first expressed, Norwegian, Swedish, and Danish were not threatened, despite the ever-growing use of English. Scientific work and publication in each national tongue had not entirely disappeared either. Rather, a certain accommodation had been reached. Preserving each national language as a form of national identity remained a theme in cultural policy (strengthened by concerns over immigration) but lacked the edge of anxiety it had attracted earlier, in part due to media comments by linguists themselves.[39] Finland, which had not engaged with such anxiety (possibly due to its longstanding use of Swedish), had moved ahead with English and a highly successful effort at internationalizing its companies and universities. Positive attitudes toward English are the rule here.[40] Sweden, dealing with immigration and a "new multilingual nation", passed a national language policy in 2005, declaring language rights to all groups and Swedish the national tongue, yet at research universities in the country, both official and unofficial policy often puts English above or on a parallel with Swedish.

With regard to scientific research itself, forms of accommodation vary. A good example is provided by the Geological Survey of Norway (NGU; http://www.ngu.no/en-gb/), a government research agency under the Ministry of Trade and Industry whose technical publications are mainly in Norwegian (~65%) but change to English when they concern areas outside of Norway and when they involve topics with international relevance or application. Geologic papers and reports in Norwegian show evidence of having adapted English terms and employed Norwegian equivalents both, so that national geologic discourse has remained fully current. Publications by university researchers in certain other fields, such as genetics and biomedicine, are exclusively in English at this point. The same is largely true for most fields in physics and mathematics. This is not surprising, since the faculty in each of these disciplines is itself international, and many research projects are now done in collaboration with teams from other countries. It would appear, therefore, that the degree of domain loss is field specific and closely related to the immediate international dimension of research.

If we ask how scientific work in Scandinavia is represented to broader domestic and international audiences, a somewhat different linguistic picture emerges. A good example here is Denmark's Agency for Science, Technology and Innovation (http://en.fi.dk/), which oversees national research activity, funding, and policy: the papers, reports, and other material it makes available are in English and Danish both if judged to be of potential international interest, but Danish only if aimed at informing Denmark citizens. The same type of approach can be found in Sweden and Norway, for example in the various publications of each country's National Research Council.

In short, domain loss in science and science-related communication has been significant but far from total. An evolving balance or semi-equilibrium seems to have developed in the use of each tongue vis-à-vis English. This leads to several questions about the long-term nature of the situation:

1. If domain loss does become complete in some fields of science for various nations, will this mean that the same will occur in other fields over time? Would it have a negative, positive, or neutral effect on the research that is done in those fields and nations?
2. Is it more helpful to see the situation of domain loss as a local problem for each national tongue or as an impact from science's own globalization?
3. In any single country, but especially a multilingual one, would linguistic diversity in science lead to greater freedom or fragmentation?
4. How much of a guide might the history of science be here? How relevant is it that the development of science in a variety of cultures—ancient Egypt, Greece, India, China, Islam, and medieval and Renaissance Europe—depended to a considerable degree on the existence of a *lingua franca*?
5. Would linguistic diversity in science guarantee diversity of thought, perspective, methods, approaches—a wider range of creative mentalities? Or rather, does arguing on behalf of linguistic diversity require us to resolve, once and for all, the Sapir-Whorf hypothesis (i.e., that language has a determining influence over thought)?

7. SUMMING UP: RISKS FROM A GLOBAL LANGUAGE

So it should be asked again whether the issues discussed here, from fairness to domain loss, are worthy of concern. The answer must be yes. Though perhaps less pernicious and widespread than claimed, they are nonetheless transitive to realities in global science today, based in English. No doubt they affect many thousands of researchers in dozens of nations.

There are indeed disadvantages to a global language for science. These include the issues of equity and fairness for those who do not possess the language, their resulting marginalization, and the incentives for finding ways of getting around the relevant barriers (e.g., plagiarism), all of which can be accentuated by bias in publication databases and related "measurements" of individual scientific standing.

The review of these issues presented in this paper, however, suggests that they are endemic to a world *lingua franca*, whether it be English, Chinese, or Swahili. Historically, moreover, English is only the latest of many international tongues for scientific work. During the modern era, most of the issues herein discussed undoubtedly existed for French (eighteenth-century Europe), German (late nineteenth and early twentieth centuries), Spanish (twentieth-century Latin America), and Russian (twentieth-century Eastern Europe). If we go back further, to Latin in medieval and Renaissance Europe and to classical Arabic in early Islam, we find all of these disadvantages very much present.[41]

What we also find, however, is that the drawbacks lessen with time. The history of *lingua franca* in science recounts, again and again, the tale of intellectual cultures adapting to new dominant tongues. Such adaptation has never come easily or quickly. It took centuries for Latin to be established throughout Western Europe and nearly 200 years for Arabic to be adopted by the intelligentsia throughout the Near East, Central Asia, and most of Persia. English, meanwhile, has been the global tongue of science for less than two generations. Given the existence of modern media, above all the internet, a much more rapid ascent has been possible for this language. For that reason alone, the disadvantages we have discussed are all the more widespread, yet may also weaken more rapidly as well.

For now and the foreseeable future, global scientific English defines a dynamic, evolving reality. The ground-level fears this generated in the 1980s and 1990s—that other national tongues in science would soon vanish—no longer seem justified. Many researchers who have not had the opportunity to learn English well continue to suffer unfortunate effects in the face of its dominance, true enough—and for these researchers, many of whom undoubtedly have valuable knowledge to share, it would obviously be best if forms of editorial help were made available. But it is also true that a great many others have learned the language well enough to move beyond this kind of situation. Younger researchers are far more likely to know English today than two or three generations ago. American and British scientists are no longer routinely over-represented on the editorial boards of international journals. Nor does non-standard English in a research paper count as a universal reason for rejection any longer, if it ever did. Scientists from the West are no longer surprised or overly impressed to encounter colleagues from Asia, Africa, or Latin America whose English is proficient, excellent, or entirely fluent. What may shock them instead is to find that far more science is published in English by foreign speakers than by Anglophones so that native speakers are less and less central to the spread of their own language. The globalization of science has brought a de-nationalization of English in research practice.

These realities therefore suggest that the issues discussed here may be temporary in importance. It argues that worldwide English capability among scientists will improve over the next few generations, unevenly and with difficulty in some countries, but progressively and also productively.

All of the drawbacks discussed in this paper, except perhaps for domain loss, concern the ability of researchers to participate more easily and fully in international science. Expansion and improvement in EFL on a global basis could well have a major positive impact over time in expanding such participation. This will not be a simple matter, given inequities among nations in educational resources, teacher availability and training, school facilities, and more. Yet the need to know English for scientific work will only grow. So much critical information and data now exists in this language, so many primary sources, international databases, archives, journals, and so forth, that its use in research has a vast momentum behind it, separate from the power and prestige of the U.S., Britain, Canada, and Australia.[42] However long English may retain its global role in science (decades? centuries?), it will remain a basic skill, an additive means of communication. Acquiring such a skill takes time, but, as the history of science in world cultures would tell us, it is what successful scientists have always done.

NOTES

1. Scott L. Montgomery, *Does Science Need a Global Language? English and the Future of Research* (Chicago: University of Chicago Press, 2013).
2. Ulrich Ammon, ed. *The Dominance of English as a Language of Science: Effects on other Languages* (Berlin: Walter de Gruter, 2001); David Crystal, *English as a Global Language* (Cambridge: Cambridge University Press, 2003); Matthew E. Falagas, Eufemia Fabritsi, Fotini C. Chelvatzoglou, and Konstantinos Rellos, "Penetration of the English Language in Science: The Case of a German National Interdisciplinary Critical Care Conference," Letter *Critical Care* 9, no. 6 (2005): 655–656; David Graddol, *English Next* (London: The British Council, 2006); Politimi Valkimadi, Drosos E. Karageorgopoulos, Harissios Vliagoftis, and Matthew E. Falagas, "Increasing Dominance of English in Publications Archived by PubMed." *Scientometrics* 81, no. 1 (2009): 219–223; Royal Society, *Knowledge, Networks and Nations: Global Scientific Collaboration in the 21st Century*. Royal Society Policy Document 03/11, March 2011. Accessed October 17, 2011. https://royalsociety.org/policy/projects/knowledge-networks-nations/report/.
3. E.g. Bo Sandelin and Nikias Sarafoglou, "Language and Scientific Publication Statistics." *Language Problems & Language Planning* 28, no. 1 (2004): 1–10.
4. Falagas, Fabritsi, Chelvatzoglou and Rellos, "Penetration of the English Language in Science: The Case of a German National Interdisciplinary Critical Care Conference"; Andreas Dinkel, Hendrik Berth, Ada Borkenhagen, and Elmar Brahler, "On Raising the International Dissemination of German Research: Does Changing Publication Language to English Attract Foreign Authors to Publish in a German Basic Psychology Research Journal?" *Experimental Psychology* 51, no. 4 (2004): 319–328.
5. Michael Nielsen, *Reinventing Discovery: The New Era of Networked Science* (Princeton: Princeton University Press, 2011).
6. Scott L. Montgomery, *Science in Translation: Movements of Knowledge in Cultures and Time* (Chicago: University of Chicago, 2001).
7. See, for example, Ammon, ed. *The Dominance of English*; Rainer Enrique Hamel, "The Dominance of English in the International Scientific Periodical

Literature and the Future of Language Use in Science." *AILA Review* (Association Internationale de Linguistique Appliquée) 20 (2007): 53–71; Humphrey Tonkin, "Language and the Ingenuity Gap." *Critical Inquiry in Language Studies* 8, no. 1 (2009): 105–116.

8. Ulrich Ammon, "Global English and the Non-Native Speaker: Overcoming Disadvantage." in *Language in the 21st Century*, ed. Humphrey Tonkin and Timothy Reagan (Amsterdam: Benjamins, 2003), 23–34; Bonnie Lee La Madeleine, "Lost in Translation." *Nature* 445 (2007): 454–455; John Flowerdew, "The Non-Anglophone Scholar at the Periphery of Scientific Communication." *AILA Review* 20 (2008), 14–27; Miguel Clavero, "'Awkward Wording. Rephrase': Linguistic Injustice in Ecological Journals." *Trends in Ecology and Evolution* 25, no. 10 (2010), 552.
9. Eugene Garfield and Alfred Welljams-Dorok, "Language Use in International Research: A Citation Analysis." *Annals of the American Academy of Political and Social Science* 511 (1990): 10–24, here 10.
10. Garfield suggested Japanese or Russian, which were logical choices in 1990. Two decades later, different choices, such as Chinese and Spanish, would make sense.
11. Education First, *EF EPI English Proficiency Index*. London: EF Education First Ltd, 2011. Accessed November 27, 2011. http://www.ef.com/epi/download-full-report/.
12. Royal Society, *Knowledge, Networks and Nations*.
13. John Flowerdew and Yongyan Li, "Language Re-use among Chinese Apprentice Scientists Writing for Publication," *Applied Linguistics* 28, no. 3 (2007): 440–465.
14. Manuel R. Guariguata, Douglas Sheil, and Daniel Murdiyarso, "'Linguistic Injustice' is not Black and White." *Trends in Ecology and Evolution* XX (2010): 1.
15. Ulrich Ammon, "Language Planning for International Scientific Communication: An Overview of Questions and Potential Solutions." *Current Issues in Language Planning* 7, no. 1 (2006): 1–30.
16. Ammon, "Global English and the Non-Native Speaker: Overcoming Disadvantage"; La Madeleine, "Lost in Translation."
17. Kerstin Stenius, Isidore Obot, Florence Kerr-Corea, Erickson F. Furtado, and Thomas F. Babor, "Reaching Parnassus: Advice on How to Get Published for Researchers from Developing or Non-English Speaking Countries," in *Publishing Addiction Science: A Guide for the Perplexed* eds. Thomas F. Babor, Kerstin Stenius, and Susan Savva (London: International Society of Addiction Journal Editors, 2004), 33–44.
18. Phyllis Freeman and Anthony Robbins, "Editorial: The Publishing Gap between Rich and Poor: the Focus of AuthorAID," *Journal of Public Health Policy* 27 (2006): 196–203.
19. See, for example, Eren Zink, *Science in Vietnam: An Assessment of IFS Grants, Young Scientists, and the Research Environment*. International Foundation for Science, MESIA (Monitoring and Evaluation System for Impact Assessment). *Impact Studies* 9 (2009). Accessed March 12, 2012. http://www.ifs.se/Publications/Mesia/MESIA_9_Vietnam.pdf.
20. La Madeleine, "Lost in Translation." *Nature* 445 (2007).
21. David Crystal, *English as a Global Language* (1997).
22. Thomson Reuters, "Science Citation Index Expanded," *Thomson Reuters*. Accessed May 5, 2012. http://thomsonreuters.com/en/products-services/scholarly-scientific-research/scholarly-search-and-discovery/science-citation-index-expanded.html.
23. Hannah Brown, "How Impact Factors Changed Medical Publishing—and Science." *British Medical Journal* 334 (2007): 561–564.
24. Eugene Garfield, "Citation Analysis as a Tool in Journal Evaluation," *Science* 178 (1972): 471–479.

25. Thed N. van Leeuwen, Henk F. Moed, Robert J.W. Tijssen, Martijn S. Visser, and Anthony F.J. van Raan, "Language Biases in the Coverage of the *Science Citation Index* and its Consequences for International Comparisons of National Research Performance." *Scientometrics* 51, no.1 (2001): 335–346; Sandelin and Sarafoglou. "Language and Scientific Publication Statistics"; Hamel, "The Dominance of English in the International Scientific Periodical Literature and the Future of Language Use in Science."
26. Peder Olesen Larsen and Markus von Ins.,"The Rate of Growth in Scientific Publication and the Decline in Coverage Provided by Science Citation Index." *Scientometrics* 84, no. 3 (2010), 575–603.
27. Bo Sandelin and Nikias Sarafoglou. "Language and Scientific Publication Statistics". *Language Problems & Language Planning* 28 (2004).
28. Netherlands Observatory of Science and Technology. *Science and Technology Indicators 2010*. Ministry of Education, Culture and Science; summary in English 2011. Accessed November 15, 2011. http://www.nowt.nl/docs/NOWT-WTI_2010_english_summary.pdf.
29. Miguel Roig, "Plagiarism in the Sciences: Conference Highlights." *Science Editor* 29, no. 2 (2006): 48–49.
30. Helen Zhang, "Chinese Journal Finds 31% of Submissions Plagiarized." *Nature* 467 (2010): 271.
31. Geoff Brumfiel, "Turkish Physicists Face Accusations of Plagiarism." *Nature* 449 (2007): 8.
32. Ihsan Yilmaz, "Plagiarism? No, We're Just Borrowing Better English." *Nature* Correspondence 449 (2007): 658
33. Flowerdew and Li, "Language Re-use among Chinese Apprentice Scientists Writing for Publication."
34. Sonia Vasconcelos, Jacqueline Leta, Lidia Costa, Andre Pinto, and Martha M. Sorenson. "Discussing Plagiarism in Latin American Science," *EMBO Reports* 10, no. 7 (2009): 677–682.
35. Scott L. Montgomery, *Science in Translation: Movements of Knowledge through Cultures and Time* (Chicago: University of Chicago Press, 2000).
36. "The appropriation of another person's ideas, data, results, or words, without giving due credit," is a general definition. But where "data" and "words" are clear, in terms of evidence of theft, "ideas" and "results" are not.
37. Tracey Bretag, "The Emperor's New Clothes: Yes, There is a Link between English Language Competence and Academic Standards." *People and Place* 15, no. 1 (2007): 13–21; Thom Bartlett and Karin Fischer. "The China Conundrum." *New York Times*, November 3, 2011.
38. Hartmut Haberland, "Domains and Domain Loss," in *The Consequences of Mobility: Linguistic and Sociocultural Contact Zones*, ed. Bent Preisler, Anne Fabricus, Hartmut Haberland, Susanne Kjaerbeck, and Karen Risager (Roskilde, DN: Roskilde University, 2005), 227–237; Bent Preisler, "Deconstructing 'the Domain of Science' as a Sociolinguistic Entity in EFL Societies: The Relationship between English and Danish in Higher Education and Research." *Ibid.*, 238–248.
39. A recent example: "English no Threat to Danish," *The Copenhagen Post*, April 20, 2011.
40. Sirpa Leppänen and 12 others. *National Survey on the English Language in Finland: Uses, Meanings, and Attitudes*, Varieng—Studies in Variation, Contacts and Change in English, Volume 5, University of Jyväskylä. Accessed October 23, 2011. https://www.jyu.fi/hum/laitokset/kielet/varieng/en/survey.
41. Scott L. Montgomery, *Science in Translation: Movements of Knowledge through Cultures and Time* (Chicago: University of Chicago Press, 2000).
42. David Graddol, *English Next* (2006).

Contributors

Jerzy Biniewicz is currently professor of University of Wrocław, working on stylistics, history of language, and textual and discursive linguistics. From 2002 to 2004 he published a number of articles concerned with pragmatic, structural, and stylistic determinants of scientific, medical, and political language, like "List Jana Pawła II (pragmatyka, struktura, język)", *Stylistyka* 16 (2007): 393–404 and "Kategoryzacja a naukowy obraz świata (słownictwo nauk ścisłych)", *Językowy obraz świata i kultura*, ed. Anna Dąbrowska, Janusz Anusiewicz (Wrocław: Wydawnictwo Uniwersytetu Wrocławskiego, 2000), as well as monographs *Rozwój polskiej terminologii chemii nieorganicznej* (Opole: Wyższa Szkoła Pedagogiczna im. Powstańców Śląskich w Opolu, 1992) and *Kształtowanie się polskiego języka nauk matematyczno-przyrodniczych* (Opole: Uniwersytet Opolski, 2003).

Matthias Dörries is a professor of history of science and director of IRIST (Institut de Recherches Interdisciplinaires sur les Sciences et la Technologie) at the University of Strasbourg. His research and interests focus on the geophysical and atmospheric sciences and the relation between science, language, and literature. He is editor of *Experimenting in Tongues* (Stanford: Stanford University Press, 2002) and of *Michael Frayn's Copenhagen in Debate* (Berkeley: Office for the History of Science and Technology, University of California Press, 2005).

Helena Durnová teaches mathematics and history of mathematics at the Faculty of Education of Masaryk University, Brno, Czech Republic. She has a MA/MSc degree in English philology and literature and in mathematics and a PhD in the history of mathematics from Masaryk University, Brno, the Czech Republic. Her research interests include history of mathematics and history of computing, in particular where communication and the use language are involved. Her recent publications include: "Embracing the Algol Effort in Czechoslovakia", *Annals of the History of Computing, IEEE Computer Society* 4.36 (2014): 26–37, and "A 'Language of Mathematics': Neutrality as a Facilitator of Circulation during the Cold War", in *The Circulation of Science and Technology: Proceedings of the 4th International Conference of the ESHS, Barcelona, 18–20 November 2010*, ed. Antoni Roca-Rosell (Barcelona: SCHCT-IEC, 2012), 1141–1146

Martin Herrnstadt is a PhD student in the history of science working group at Goethe University Frankfurt am Main. Since 2012 he has been working on a project with Laurens Schlicht and Moritz Epple investigating the epistemological transformation of the *science de l'homme* around 1800. He is currently working on the history of the sciences of society and administration in postrevolutionary and *restauration* France. His recent publication "Vom Geheimnis des Staates zum Staat als Geheimnis. Aspekte der Geschichte der Statistik in Frankreich (ca.1661–1800)" deals with the intertwined histories of statistics and secrecy (published in: *Geheimnis_Wissen. Perspektiven auf das Wissen vom Geheimnis seit dem 18. Jahrhundert*, eds. Laurens Schlicht and Sebastian Klinge [Berlin: Trafo, 2014, 65–94]).

Markus Krajewski is professor of media history at the University of Basel, Switzerland. Among his books are *Paper Machines. About Cards & Catalogs, 1548–1929* (Cambridge, MA: The MIT Press, 2011); *Der Diener. Mediengeschichte einer Figur zwischen König und Klient* (Frankfurt am Main: S. Fischer, 2010), translated now into English as *The Servant. Media History of a Figure between King and Client* (Yale University Press, forthcoming); and *World Projects. Global Information Before WWI* (Minneapolis: University of Minnesota Press, 2014). His current research projects include the problem of planned obsolescence, media and architecture, epistemology of the peripheral, and the history of exactitude in scholarly and scientific contexts. He is also developer and maintainer of the bibliography software *synapsen—a hypertextual card index* (www.verzetteln.de/synapsen). For further information see: gtm.mewi.unibas.ch.

Miles MacLeod is assistant professor of philosophy of science at the University of Twente, The Netherlands. He has a doctoral degree in history and philosophy of science from the University of Vienna. He works and publishes on themes connected to model-building practices in the biological sciences, scientific cognition, and interdisciplinary collaboration. He has a strong interest in the historical development of science and its relevance for understanding modern-day practices. Key publications include, "Building Simulations from the Ground-Up: Modeling and Theory in Systems Biology", *Philosophy of Science* 80.4 (2013): 533–556, with Nancy Nersessian, and "Rethinking Scientific Concepts for Research Contexts: The Case of the Classical Gene", in *Scientific Concepts and Investigative Practice*, eds. Friedrich Steinle and Uljana Feest (Berlin: De Gruyter, 2012).

Scott L. Montgomery is an affiliate faculty member in the Henry M. Jackson School of International Studies, University of Washington, Seattle (USA). He has written widely on topics related to the history of science, language studies, translation, and English as a global language. He has an abiding interest in the history of lingua franca in science and in premodern scientific cultures, as well as in the role of scientific ideas in the construction of modernity. Such interests are shown by several of his books: *A History of Science in World Cultures* (Routledge, 2015); *The Shape of the New: Four Big Ideas and How They Built the Modern World* (Princeton, 2015); *Does Science Need a Global Language? English and the Future of Research* (Chicago, 2013); and *Science in Translation: Movements of Knowledge in Cultures and Time* (Chicago, 2001).

Josefina Rodríguez Arribas is currently fellow at the IKGF-Internationales Kolleg für Geisteswissenschaftliche Forschung Friedrich-Alexander-Universität-Erlangen-Nürnberg, Germany. She has substantial research experience in the field of the history of medieval science (astronomy, astrology, and scientific instruments) and medieval Hebrew language, particularly the emergence and constitution of technical terminology in medieval Hebrew. She is especially interested in critical approaches to the definition of pre-modern science, the presence of science in non-scientific texts and contexts, and the relations of textual and material cultures in medieval science. She has worked on the transmission of Greek and Roman knowledge to Europe through Arabic and Hebrew texts and translations. She also has long experience with medieval manuscripts and museum objects. Since 2009, her research has focused on the relations between the textual and material scientific cultures of Jews, notably the cultural role of astrolabes among Jews in Europe and the Near East between the 12th and 17th centuries. She has published the book *El cielo de Sefarad: los judíos y los Astros* (Córdoba: El Almendro, 2011), and several chapters and articles on medieval astronomy and astrology, and is preparing two books on the role of astrolabes in medieval Jewish culture (texts and instruments), which include critical editions of the Hebrew treatises and a catalogue of astrolabes with Hebrew script.

Laurens Schlicht is researcher in a research group working on the history of the sciences of man at Goethe-University Frankfurt am Main (wg.geschichte.uni-frankfurt.de/epi-m/). Together with his colleague Martin Herrnstadt he focuses on the *Société des observateurs de l'homme* (Society of the Observers of Man) from the viewpoint of historical epistemology. His recent publications include: "Geheimnis und Unendlichkeit bei Cureau de la Chambre und Concordet", in *Was als wissenschaftlich gelten darf. Praktiken und Grenzziehungen in Gelehrtenmilieus der Vormoderne*, eds. Martin Mulsow and Frank Rexhoth (Frankfurt am Main: Campus, 2014, 356–387); ed. with Sebastian Klinge, *Geheimnis_Wissen.*

Perspektiven auf das Wissen vom Geheimnis seit dem 18. Jahrhundert (Berlin: trafo 2014).

Ekaterina Smirnova is currently affiliated with Sciences Po (Paris) and the STS Center in European University at St. Petersburg (EUSP) in the framework of the research project "Russian Computer Scientists at Home and Abroad." She has a PhD in cultural anthropology from St. Petersburg State University. She holds an MA in sociology and social anthropology from Central European University, Budapest, and MA in sociology from EUSP. Among her research interests are conceptual history, history of science in the early modern period, theory of republicanism, and science and technology studies.

Rocío G. Sumillera is Assistant Professor in the Department of English and German, University of Granada, Spain. Her research interests revolve around history of translation and early modern rhetoric and poetics. She is co-editor of the volume *The Failed Text: Literature and Failure* (Newcastle: Cambridge Scholars Publishing, 2013) and author of the critical edition of Richard Carew's *The Examination of Mens Wits* (1594), the first translation into English of Juan Huarte de San Juan's *Examen de ingenios para las ciencias* (1575) (London: Modern Humanities Research Association, 2014). She is currently working on a translation into Spanish of John Knox's *The First Blast of the Trumpet against the Monstruous Regiment of Women* (1558), to be published by Tirant lo Blanch.

Jan Surman is postdoctoral research associate at the Leibniz Graduate School History, Knowledge, Media in East Central Europe at the Herder Institute for Historical Research on East Central Europe in Marburg. He received his PhD in history from the University of Vienna in 2012. He studied in the PhD program "The Sciences in Historical Context" at the University of Vienna. He has been a visiting scholar at the Max Planck Institute for History of Science, Berlin; the Center for Austrian Studies, University of Minnesota; the Institute for the History of Science of the Polish Academy of Sciences, Warsaw; the Centre for the History of Sciences and Humanities, Prague; and the Department of History, Princeton University. His publications include, ed. with Mitchell G. Ash, *The Nationalization of Scientific Knowledge in the Habsburg Empire, 1848–1918* (Basingstoke, UK: Palgrave Macmillan, 2012); eds. with Katalin Straner und Peter Haslinger, "Concepts in Focus: Nomadic Concepts", *Contributions to the History of Concepts* 9.2 (2014); "The Circulation of Scientific Knowledge in the Late Habsburg Monarchy: Multicultural Perspectives on Imperial Scholarship", *Austrian History Yearbook* 46 (2015): 163–182.

Priya Venkatesan Hays, PhD, received her AB from Dartmouth College, her MS in genetics from the University of California, Davis, and her PhD in literature from the University of California, San Diego. A prolific science writer, she has over 20 publications on science, technology, and society in peer-reviewed journals and has presented her papers at over 30 conferences in the US and abroad. Her most recent books have included *Science, Cultural Values and Ethics* (Champaign, IL: Common Ground Publishing, 2013) and *Molecular Biology in Narrative Form* (New York: Peter Lang Publishing, 2006). She has taught writing in the sciences at U.C. Berkeley, Santa Clara University, and San Francisco State University and completed an interdisciplinary postdoctoral fellowship at Dartmouth Medical School.

Index

For Product Safety Concerns and Information please contact our EU
representative GPSR@taylorandfrancis.com
Taylor & Francis Verlag GmbH, Kaufingerstraße 24, 80331 München, Germany

www.ingramcontent.com/pod-product-compliance
Ingram Content Group UK Ltd.
Pitfield, Milton Keynes, MK11 3LW, UK
UKHW022000180926
475701UK00003B/120
* 9 7 8 0 3 6 7 2 6 3 9 2 8 *